Additive Manufacturing of Polymers for Tissue Engineering

Application of additive manufacturing and tissue engineering in the fields of science and technology enables the manufacturing of biocompatible, customized, reliable, and cost-effective parts, restoring the functionality of a failed human body part. This book offers a platform for recent breakthroughs in additive manufacturing related to biomedical applications.

This book highlights some of the top innovations and advances in additive manufacturing and processing technologies that are the future of the manufacturing industry while also presenting current challenges and opportunities regarding the choice of material. This book includes areas of applications such as surgical guides, tissue regeneration, artificial scaffolds, implants, and drug delivery and release. Throughout the book, an emphasis is placed on rapid tooling for engineering applications.

Additive Manufacturing of Polymers for Tissue Engineering: Fundamentals, Applications, and Future Advancements acts as a first-hand source of information for academic scholars and commercial manufacturers as they make strategic manufacturing and development plans.

Sustainable Manufacturing Technologies: Additive, Subtractive, and Hybrid

Series Editors: Chander Prakash, Sunpreet Singh, Seeram Ramakrishna, and Linda Yongling Wu

This book series offers the reader comprehensive insights of recent research breakthroughs in additive, subtractive, and hybrid technologies while emphasizing their sustainability aspects. Sustainability has become an integral part of all manufacturing enterprises providing various techno-social pathways toward developing environmentally friendly manufacturing practices. It has also been found that numerous manufacturing firms are still reluctant to upgrade their conventional practices to sophisticated sustainable approaches. Therefore, this new book series is aimed to provide a globalized platform to share innovative manufacturing mythologies and technologies. This book will help the manufacturers to fabricate the complicated structure with better properties that is the most difficult task of conventional and nonconventional manufacturing technologies and cover recent innovations.

Advances in Manufacturing Technology
Computational Materials Processing and Characterization
Edited by Rupinder Singh, Sukhdeep Singh Dhami, and B. S. Pabla

Additive Manufacturing for Plastic Recycling
Efforts in Boosting A Circular Economy
Edited by Rupinder Singh and Ranvijay Kumar

Additive Manufacturing Processes in Biomedical Engineering
Advanced Fabrication Methods and Rapid Tooling Techniques
Edited by Atul Babbar, Ankit Sharma, Vivek Jain, and Dheeraj Gupta

Additive Manufacturing of Polymers for Tissue Engineering
Fundamentals, Applications, and Future Advancements
Edited by Atul Babbar, Ranvijay Kumar, Vikas Dhawan, Nishant Ranjan, and Ankit Sharma

For more information on this series, please visit: https://www.routledge.com/ Sustainable-Manufacturing-Technologies-Additive-Subtractive-and-Hybrid/ book-series/CRCSMTASH

Additive Manufacturing of Polymers for Tissue Engineering

Fundamentals, Applications, and Future Advancements

Edited by
Atul Babbar
Ranvijay Kumar
Vikas Dhawan
Nishant Ranjan
Ankit Sharma

CRC Press
Taylor & Francis Group
Boca Raton London New York

CRC Press is an imprint of the
Taylor & Francis Group, an **informa** business

First edition published 2023
by CRC Press
6000 Broken Sound Parkway NW, Suite 300, Boca Raton, FL 33487-2742

and by CRC Press
4 Park Square, Milton Park, Abingdon, Oxon, OX14 4RN

CRC Press is an imprint of Taylor & Francis Group, LLC

© 2023 Taylor & Francis Group, LLC

Library of Congress Cataloging-in-Publication Data
Names: Babbar, Atul, editor. | Kumar, Ranvijay, editor. | Dhawan, Vikas, editor. | Ranjan, Nishant, editor. | Sharma, Ankit (Professor of mechanical engineering), editor.
Title: Additive manufacturing of polymers for tissue engineering : fundamentals, applications, and future advancements / edited by Atul Babbar, Ranvijay Kumar, Vikas Dhawan, Nishant Ranjan, Ankit Sharma.
Description: First edition. | Boca Raton : CRC Press, 2023. | Includes bibliographical references and index. | Summary: "The book highlights some of the top innovations and advances in additive manufacturing along with processing technologies. It captures how 3D printing has distinct advantages in improving quality, cost-effectiveness, and has higher efficiency compared to traditional manufacturing processes, Current challenges and opportunities regarding choice of material, design, and efficiency are presented and a discussion on selected areas of application such as surgical guides, tissue regeneration, artificial scaffolds, implants, and drug delivery and release is provided. An emphasis throughout the book is on rapid tooling for engineering applications"-- Provided by publisher.
Identifiers: LCCN 2022014823 (print) | LCCN 2022014824 (ebook) | ISBN 9781032210421 (hardback) | ISBN 9781032210452 (paperback) | ISBN 9781003266464 (ebook)
Subjects: LCSH: Tissue engineering. | Polymers in medicine. | Additive manufacturing.
Classification: LCC R857.T55 A33 2023 (print) | LCC R857.T55 (ebook) | DDC 610.28--dc23/eng/20220628
LC record available at https://lccn.loc.gov/2022014823
LC ebook record available at https://lccn.loc.gov/2022014824

ISBN: 978-1-032-21042-1 (hbk)
ISBN: 978-1-032-21045-2 (pbk)
ISBN: 978-1-003-26646-4 (ebk)

DOI: 10.1201/9781003266464

Typeset in Times
by SPi Technologies India Pvt Ltd (Straive)

Contents

Editors

Dr. Atul Babbar is an assistant professor at the Mechanical Engineering Department of SGT University, Gurugram. He has been teaching students in academics and research fields. His research interests include biomedical machining, 3D and 4D printing, and conventional/non-conventional machining. He has authored several research articles and book chapters that are published in various international/national Web of Science and Scopus journals. He has been granted numerous national and international patents. He has been reviewing research articles of various peer-reviewed SCI and Scopus indexed journals.

Dr. Nishant Ranjan is an assistant professor at University Centre for Research and Development in the Chandigarh University. Fused deposition-modeling, extrusion, thermoplastic polymers, composition of thermoplastic polymers, natural and synthetic biopolymers, scaffolds printing, 3D printing technology, thermal, mechanical, morphological, and chemical properties of thermoplastic polymers, biocompatible and biodegradable fillers, and reinforcement of materials are the main focus areas of Dr. Nishant Ranjan. He has coauthored more than 12 research papers in science citation index journals and 22 book chapters and has presented ten research papers at various national-level conferences.

Dr. Ranvijay Kumar is an assistant professor in University Center for Research and Development, Chandigarh University. He has received PhD in Mechanical Engineering from Punjabi University, Patiala. His research interests include additive manufacturing, shape memory polymers, smart materials, friction-based welding techniques, advance materials processing, polymer matrix composite preparations, reinforced polymer composites for 3D printing, plastic solid waste management, thermosetting recycling, and destructive testing of materials. He has co-authored more than 55 research papers in science citation index journals and 35 book chapters and has presented 20 research papers at various national/international-level conferences.

Dr. Vikas Dhawan is an academician and administrator in the field of higher education with more than 22 years of rich experience. He has more than 12 years of administrative experience as a principal, additional director, and director & director principal in various reputed institutions. He has filed five patents and has more than 40 research publications in international and national journals and conferences of repute. He has organized many conferences, seminars, fused deposition modeling, and workshops and has chaired sessions in international and national conferences. He has established many labs, incubation centers, and centers of excellence in the field of mechanical engineering.

Dr. Ankit Sharma is currently working as an assistant professor in the Department of Mechanical Engineering, Chitkara College of Applied Engineering at Chitkara University, Punjab, India. He completed his doctoral degree from Thapar Institute of Engineering and Technology, India. He has authored numerous national and international publications in SCI, Scopus, and Web of Science indexed journal. He has filed/published 20+ national/international patents. His research interests include 3D printing, modern machining, machining of hard and brittle materials, and additive manufacturing. He also has vast experience in industry as well as in academics.

Contributors

Atul Babbar
Shree Guru Gobind Singh
 Tricentenary University,
Gurugram, India

Kannan Ganesa Balamurugan
IFET College of Engineering
Valavanur, Villupuram, Tamil Nadu, India

Jasgurpreet Singh Chohan
Chandigarh University
Mohali, India

Lalita Chopra
Chandigarh University
Mohali, India

Dharmpal Deepak
Punjabi University
Patiala, India

Sulakshna Dwivedi
Punjabi University
Patiala, India

Sehra Farooq
Chandigarh University
Mohali, India

Harnam Singh Farwaha
Guru Nanak Dev Engineering College
Ludhiana, India

Jaspreet Kaur
Punjabi University
Patiala, India

Swapandeep Kaur
Department of Electrical Engineering
Guru Nanak Dev Engineering College
Ludhiana, Punjab, India

Raman Kumar
Department of Mechanical and
 Production Engineering
Guru Nanak Dev Engineering College
Ludhiana, Punjab, India

Ranvijay Kumar
Chandigarh University
Mohali, India

Vidyapati Kumar
Indian Institute of Technology
Kharagpur, India

Manikanika
Chandigarh University
Mohali, India

Jasmine Nindra
Shree Guru Gobind Singh Tricentenary
 University
Gurugram, India

Mona Prabhakar
Shree Guru Gobind Singh Tricentenary
 University
Gurugram, India

G. Prabu
National Institute of Technology
Tiruchirappalli, India

Nishant Ranjan
University Centre for Research and
 Development
Chandigarh University, Mohali, India

Ankit Sharma
Chitkara University
Rajpura, India

Gurpreet Singh
Punjabi University
Patiala, India

Sandeep Singh
Punjabi University
Patiala, India

Tapinderjit Singh
Punjabi University
Patiala, India

Yebing Tian
Shandong University of Technology
Zibo, People's Republic of China

Ankit Tyagi
Shree Guru Gobind Singh Tricentenary
   University
Gurugram, India

1 3D Bioprinting in Biomedical Applications

Atul Babbar
Shree Guru Gobind Singh Tricentenary University,
Gurugram, India

Yebing Tian
Shandong University of Technology, Zibo,
People's Republic of China

Vidyapati Kumar
Indian Institute of Technology, Kharagpur, India

Ankit Sharma
Chitkara University, Rajpura, India

CONTENTS

1.1 INTRODUCTION TO 3D BIOPRINTING PROCESS

Three-dimensional (3D) bioprinting has evolved as a fairly simple approach for producing the prototype using bioinks in a precise form of the organ such that the cell grows in that printed component and subsequently develops as a fully functioning organ. It has acquired widespread adoption because it is suitable for producing prototypes for visual inspection purposes and because of its advantages such as great flexibility, customization depending on patients, scalability, reliability, durability, and extremely high speed. Three-dimensional bioprinting is widely used due to these characteristics in the biomedical area and allows for more design and complexity. Bioprinting is often used to print tissues and organs suitable for transplantation by integrating scaffolds, living cells, and numerous other biological bioactivated agents. These capabilities for tissue regeneration bring up a broad range of possibilities in the field of tissue engineering. There are numerous advanced manufacturing processes that are often used in this field to understand the behavior of the machining operation, which aids in the future use of those processes in the production of biomedical devices(Babbar et al., 2019b; Babbar et al., 2020; Baraiya et al., 2020; Sharma et al., 2020).

Advanced 3D bioprinting technology has acquired widespread appeal in the current environment due to its several advantages over traditional and nontraditional manufacturing processes, including cost reduction, speed, design flexibility, one-step manufacture, and sustainable development. Three-dimensional bioprinting is the method of creating a 3D object layer by layer out of a computer-aided design (CAD) which streamlines the machining procedure and eliminates extraneous material out of a larger stock. This technique is unusual because it immediately converts the material to its final shape with minimal material waste while achieving the requisite geometric proportions. Due to the presence of a more extensive spectrum of 3D bioprinting processes (Bose et al., 2018; Herzog et al., 2016; Labonnote et al., 2016; Wang et al., 2017), this additive manufacturing process is used for a specific material and application, making it a challenging effort to determine the appropriate additive manufacturing technique. Different methods offer different surface finishes, high-dimensional precision, and post-processing needs depending on the material used for printing. Additive manufacturing has applications in various fields, including aerospace, structural, biomedical, and complicated component production, among others (Sharma et al., 2018).

Furthermore, 3D bioprinting has piqued the curiosity of many experts due to its adaptability in producing low-cost, quick surgical equipment, and patient-specific bio-implants (Chakraborty et al., 2018; Kumar & Chakraborty, 2022; Prakash et al., 2021; Sharma et al., 2021; Zuback & DebRoy, 2018). Three-dimensional bioprinting has been used successfully to repair or produce anatomical tissues such as bioartificial livers and bioartificial cardiovascular systems; however, it is not restricted to orthopedic implants and may also be used to build medical electronic microfluidic devices. Stress shielding occurs in this circumstance due to a mechanical mismatch between metallic implants and bodily organs, resulting in organ and implant failure. This form of implant failure is more common in implants that are made using traditional manufacturing techniques.

The 3D bioprinting technique attempts to produce repeatable, sophisticated tissue frameworks that are efficiently vascularized and suited for therapeutical usage. Since biological organ systems comprise unstructured 3D shapes composed of many types of cells and extracellular matrix (ECM) with biological structure, this approach may be the more viable solution to attain this objective (Labonnote et al., 2016). A standard approach for 3D bioprinting begins with the production of 3D volumetric information generated by scanning the patient with different imaging equipment such as computed tomography and magnetic resonance imaging. The next step is to gather these data using sectional segments of the human body and save it in the DICOM

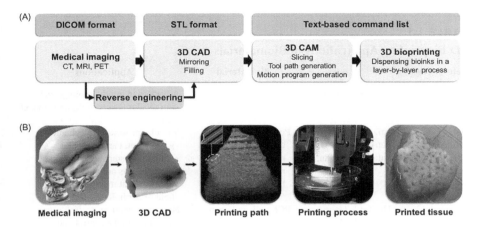

FIGURE 1.1 Process of 3D bioprinting (Huh et al., 2020).

(Digital Imaging and Communications in Medicine) format. The next step is to use reverse engineering to convert this information into a 3D model. Furthermore, the reconstruction of the complete CAD model is built by extracting localized volumetric data from the tissue framework and modeling its surface, since the tissues and organs present in the human body are quite detailed, which may impact the bioprinting process. Following that, a computer-aided manufacturing (CAM) system is used in three phases: slicing, tool path construction, and motion program development to print and execute the predetermined route. As a result, a well-planned approach for tool path creation is essential. Figure 1.1 depicts a 3D bioprinting technique that begins with a medical picture and ends with printed tissue constructions created using the CAD or CAM technology and automated printing of 3D shapes that mimic target tissue or organ (Huh et al., 2020).

1.2 APPLICATION OF BIOMATERIALS IN 3D BIOPRINTING PROCESS

Biomaterials have been used in regulated pharmaceutical delivery techniques, sutures and adhesives, cardiac bypass, rehabilitative and orthopedic devices, ocular devices including corneas and corrective lenses, and dentistry (Babbar, Jain, Gupta, Prakash, et al., 2020e). Several experiments have been carried out in the field of biomedical engineering, such as neurosurgical bone grinding, in which multi-criteria decision-making and optimization tools have been successfully employed to fine tune the parametric setting, which aids in the 3D bioprinting process (Babbar et al., 2021; Babbar, Jain, et al., 2019d; Babbar, Jain, Gupta, & Agrawal, 2021e; Chakraborty et al., 2019; Tappa & Jammalamadaka, 2018). Several components, including titanium, have been used to generate therapeutic implants. Bioceramics, polymers, metals, and composites are just a handful of the materials accessible. Table 1.1 outlines the current biological applications for biomaterials.

TABLE 1.1

3D Bioprinting Applications of Biomaterials

References	Biomaterial	Applications
ASM International (2003); Babbar, Jain, & Gupta (2020c)	Silicones, hydrogels	Advances in ophthalmology such as intraocular lenses (IOL) and contact lenses.
Babbar et al. (2021a); Chakraborty & Kumar (2021); Yuan et al. (2004); Zdrahala & Zdrahala (1999); Zhu et al. (2004)	Polymeric component, metallic material, ceramic material, PU, PE, PBE, PP, SS	Vascular prosthesis, mechanical gates, stent, cardiovascular pumping systems, blood bags, and catheters are all examples of products that may be used in the field of cardiology.
Cao et al. (2013); Horner & Gage (2000); Kumar et al. (2020b); Kumar et al. (2020a); Scheib & Höke (2013); Stratton et al. (2016)	PCL, silk, collagen	Nerve regeneration scaffolds are used in both the central and peripheral nervous systems.
Jin et al. (2021); Kadambi et al. (2021); Karst & Yang (2006); Liu et al. (2021); Stratton et al. (2016); Tan et al., (2020)	PLLA, PDLA	Medication transmission, vascular bypass, skeletal fasteners, stabilization PEGs, and dermal filler for facial atrophy are all used in the treatment of facial atrophy.
Babbar, Jain, et al. (2019c); Tan et al. (2020)	PLGA	Pharmaceutical delivery
Jędzierowska et al. (2021); Stratton et al. (2016)	PCL	Everlasting implantation; medication release; maxillo-cranial facial implant
Dong et al. (2021); Hernandez et al. (2021); Hu et al. (2021)	PCL-gelatin, PCL-chitosan, PCL-collagen	Rejuvenation of tissues
Pérez-Merino et al. (2013); Terrada et al. (2011)	PPD	Intrinsic rupture fixation, medical implants in the form of films, foaming agents, and molded scaffold are all possible applications.
Babbar et al. (2019); Teo et al. (2016); Zhou et al. (2014)	LDPE, HDPE	Rhinoplasty surgeries, comprehensive hip arthroplasty, and osteolysis therapy using polymer–ceramic composites
Gao et al. (2017); Peltola et al. (2012)	PMMA	Dentistry implantation for rehabilitation and aesthetics, orbital medicinal implantation, rhinoplasty, cranioplasty, bone cement in hip joint restoration.
Babbar, Jain, Gupta, Prakash, et al. (2020c); Kumar et al. (2021); Tayyaba et al. (2021)	PDMS	Implantable electrical equipment and sensors, clinical implants, esophageal analogues, catheters, shunts, blood pumps, and pacemakers are all included inside this enclosure.
Chakraborty et al. (2017); Kubyshkina et al. (2011); Singh et al. (2019)	PA such as nylon and nylon-composites	Sutures, denture manufacture, bone regeneration scaffold material, and nanofillers

(Continued)

TABLE 1.1 (*Continued*)
3D Bioprinting Applications of Biomaterials

References	Biomaterial	Applications
Babbar, Jain, Gupta, Singh, et al. (2020a); Li et al. (2011)	CNT and composite material	Metallic sealants for musculoskeletal implantable devices to increase permeability of the substrate, minimize metallic ionization, and enhance hydroxyapatite production

Note: Polyurethane (PU), Polyesters (PE), Polybutesters (PBE), Polypropylene (PP), Stainless steel (SS), Polycaprolactone (PCL), Poly(l-lactic acid) (PLLA), Poly(lactic acid), Poly(lactic-co-glycolic acid) (PLGA), Polycaprolactone (PCL), poly(d-lactic acid) (PDLA), Poly-para-dioxanone (PPD), Low-density polyethylene (LDPE), High-density polyethylene (HDPE), Poly(methyl methacrylate) (PMMA), Polydimethylsiloxane (PDMS), Polyamides (PA), Polyethylene glycols (PEGs), Carbon nanotubes (CNT), and composites.

1.3 3D BIOPRINTING TECHNIQUES

Several 3D printing technologies have been created to bioengineer 3D tissue or organ frameworks for biomedical purposes, as demonstrated in Figure 1.2. The most extensively used forms of 3D bioprinting processes are extrusion-based bioprinting, laser-assisted bioprinting, and laser-based stereolithography. The effectiveness

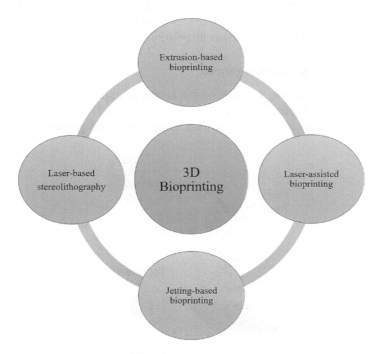

FIGURE 1.2 Types of 3D bioprinting.

of each printing technique is heavily reliant on biomaterial choices and functions (Babbar, Jain, Gupta, Agrawal 2021d).

1.3.1 JETTING-BASED BIOPRINTING

Jetting-based bioprinting is the earliest printing method, in which bioinks are used to print. These bioinks may be either natural or manufactured substances that help in cell adherence, propagation, and replication. Bioink is pushed with force via a nozzle in this approach, resulting in a spray of droplets. These printers may have a single or several print heads. A chamber and a nozzle are both included in each print head. The surface tension of the fluid keeps the bioinks near the nozzle opening. In three ways, pressure pulses are injected into the print head chamber. As seen in Figure 1.3, it is provided via piezoelectric inkjet, thermal inkjet, or electrostatic bioprinting. The actuator in the piezoelectric inkjet generates pressure pulses to deposit the bioinks; however, certain print heads need back pressure to complement the pressure pulses to make droplets of bioinks. When a voltage pulse is supplied to a thermal inkjet printer's thermal actuator, it locally warms the bioink solution. Figure 1.3 (Huh et al., 2020) shows that local heating produces a vapor bubble. This bubble rapidly expands and shrinks, generating a force burst inside the fluid compartment and driving the bioink droplet to defy interfacial tension and accumulate on the scaffold. Thermal inkjet printers may discharge biological materials such as proteins and mammalian cells, among other things. Bioink droplets are created in electrostatic bioprinters by increasing the capacity of the fluid compartment with the aid of a bioink fluid attached to the plate. After that, the pressure plate deflects between the electrode and the plate once the voltage is applied. Finally, as the voltage drops, the bioink is evacuated as the pressure plate re-establishes its position, and printing happens.

1.3.2 EXTRUSION-BASED BIOPRINTING

Extrusion-based bioprinting is based on the notion of applying extrusion pressure to the bioink, which is beneficial for tissue regeneration and repair. The bioink contained in this process is largely deposited using pneumatic pressure, mechanical

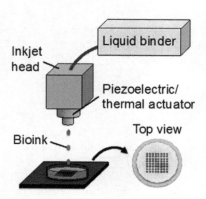

FIGURE 1.3   Jetting-based 3D bioprinting (Huh et al., 2020).

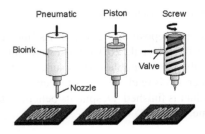

FIGURE 1.4 Extrusion-based 3D bioprinting (Huh et al., 2020).

pressure in the form of a screw or piston, and lastly the substrate is extruded out, as illustrated in Figure 1.4 (Huh et al., 2020). The robotic stage controller governs and controls the whole extrusion process of the bioprinter. The head can move in three directions: x, y, and z, and the bioink may be dispensed directly onto the substrate underneath it. These printers are capable of dispensing high cell density bioinks, unique hydrogels with a wide range of viscosities, and biodegradable thermoplastics like polycaprolactone. When compared with inkjet printers, extrusion bioprinting reduces the possibility of bioink clogging. The main disadvantage of extrusion is that we must guarantee that the shear force is not so great that it impairs cell viability.

1.3.3 Laser-Assisted Bioprinting

Laser-assisted bioprinting is another popular method for bioprinting live cells onto a substrate. This printing is made feasible by using a high-intensity light source or a light with a long wavelength. A laser pulse, focusing lens, donor slide, energy absorption layer, donor substrate, and collector slide are the main components of a laser bioprinter, as depicted in Figure 1.5. The focusing lens in Figure 1.5 concentrates the high intensity light, after which the bioink is concentrated on the collection slide and the printed output is formed. Unlike inkjet printers, laser printers have no nozzles and may therefore deposit large densities of bioinks without clogging, as demonstrated in Figure 1.5 (Huh et al., 2020).

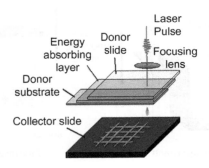

FIGURE 1.5 Laser-assisted bioprinting (Huh et al., 2020).

1.3.4 Laser-Based Stereolithography

It is a free form process for depositing light to cross-linked polymer materials, as demonstrated in Figure 1.6 (Huh et al., 2020). Most stereolithography techniques use UV light, directed onto the photocurable resin's surface. When the resin has dried, the platform travels higher and is ready to deposit a new coat of resin. This procedure is repeated until the product is finished. Despite its benefits, stereolithography has an extended processing time.

1.3.5 Biomaterials as Bioinks for 3D Bioprinting

Bioinks are fluid compositions that comprise three or four matrix constituents and are supplied into a bioprinter before being accumulated on the scaffold. Such scaffolds allow cells to connect, survive, bioaccumulate, and replicate after printing (Babbar et al., 2021). This cell proliferation phenomenon is important in tissue regeneration because it assists in the rehabilitation of dead body tissues. These multicomponent bioinks are used to produce various tissue constructions. The reason for employing multicomponent bioinks is because natural polymers such as gelatin and collagen are extensively utilized bioinks that aid in cell adhesion and migration. However, they have limited mechanical capabilities; to overcome these limitations, various additives and biomaterials are combined to make a multicomponent biomaterial with enhanced qualities. Rheological characteristics of multicomponent bioinks are critical; they must be accurately managed in order to provide optimum printability and structural stability of the structure.

Multiple bioinks may be combined in multicomponent bioinks to provide varied stiffness. These bioinks, which are constructed of cross-linked polymers, may solidify immediately after bioprinting. If they harden too soon, they may jam the printer nozzle. If it solidifies too slowly, the structure will collapse. These bioink multimaterial combinations should not be hazardous in the short or long term (Zhao et al., 2015). Bioinks are further subdivided into shear thinning and shear thickening materials (Guzzi et al., 2021). Shear thinning materials allow printers to avoid using shear force, and they restore their shape after shear force is removed. Gelation occurs in

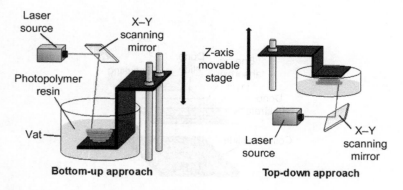

FIGURE 1.6   Laser-based stereolithography (Huh et al., 2020).

TABLE 1.2
Application of Bioinks in 3D Bioprinting

Molecular Type	3D Bioprinting Technique	Features
Collagen – It serves as the fundamental structural component of skin-associated tissues.	Pressure based	It has significant biological importance (Huh et al. 2020).
Elastin – It has a high degree of hydrophobicity.	Pressure based	It has a potent biological impact (Zhao et al. 2015).
Fibrin – It is created by polymerizing fibrinogen, a soluble plasma protein.	Extrusion-based bioprinting (Mechanical/Pneumatic	It exhibits quick gelation (Kolesky et al. 2014).
Gelatin – It is a protein molecule derived from the partial hydrolysis of collagen.	Coaxial extrusion printing	It possesses excellent biocompatibility, rapid liquid solubility, and thermally reversible gelation (Guzzi et al. 2021).
Agarose – Extracted from seaweed	Pressure based	It possesses great stability, effective crosslinking, and is nontoxic (Norotte et al. 2009).
Alginate – Natural bioink	Coaxial extrusion printing	It exhibits mild cross-linking characteristics, rapid gelation, and great biocompatibility (Babbar, Jain, Gupta, & Agrawal 2021d).
Chitosan – Natural bioink	Coaxial extrusion printing	It has excellent biocompatibility and antimicrobial characteristics (Hu et al. 2021).
Silk fibroin – Natural bioink	Stereolithography	It is biologically relevant and gels quickly (Babbar, Jain, et al. 2019c).

shear thickening materials in response to chemical and physical stimuli. When exposed to UV radiation, chemical cross-linking occurs.

The same biomaterial bioink cannot be used to print all tissues and organs because it may not meet the mechanical and functional needs of tissues or organs such as multiplication and dissemination. Some biomaterials, like as polyethylene glycol, have variable molecular weight and may join printed tissue, but they do not proliferate. Gelatin and fibrin, for example, are examples of natural biomaterials with poor mechanical characteristics. As a result, hybrid bioinks are created by combining more than one bioink, which is responsible for improving component quality as well as providing an environment conducive to cell growth, and these hybrid bioinks have acquired widespread acceptance. Table 1.2 lists some of the bioinks that may be combined to create a hybrid bioink.

1.4 THREE-DIMENSIONAL BIOPRINTING IN TISSUE ENGINEERING APPLICATIONS

The absence of appropriate vasculature into implantable frameworks is a significant restriction in bioengineered tissue constructs. Furthermore, highly vascularized tissue constructions are required for the proper functioning and survival of tissues and cells as a 3D tissues framework packed from metabolic active cells that are account-able for the establishment of necrotic cores in the absence of the vascular struc-ture because it is limited to transporting nutrients toward and away from the tissues (Huang et al., 2015; Mondy et al., 2009). When a 3D tissue framework containing cells is implanted, the successful transmit of mass necessitates an intact microvas-cular network for sustaining the metabolically active activities of the cells inside the framework. The ingrowth of the microvascular network into the implanted bioengi-neered tissue framework, on the other hand, happens in a timely way, which is the most important accomplishment for therapeutic reasons (Murphy & Atala, 2014). As a result, some researchers have acquired an interest and sought to manufacture and build the vascularized tissue framework using 3D bioprinting methods, which is a potential approach for this sort of application (Huh et al., 2020).

The simplest and most fundamental method of fabricating a vascular-shaped tis-sues framework is to use a sacrificial component that functions like a structure during printing and is then removed to leave a hollow-shaped tunnel-like structure. Miller et al. (2012) investigated a similar kind of work in which the vascular casting process was used using carbohydrate glass as a sacrificial template. The initial stage in this method is to inject a cell-based loaded hydrogel into a mold with a carbohydrate lat-tice structure, which is then removed once the hydrogel has been cross-linked. Human umbilical vein endothelial cells (HUVECs) are then injected into the lumen that has been created in an attempt to develop a vascular structure. Furthermore, extrusion-based bioprinting has been used successfully in this scenario to generate the vascular structure. Kolesky et al. (2014) proposed an approach for fabricating a 3D framework with vasculature that includes many kinds of cells and ECM proteins, using coprinting of two bioinks of HUVEC (red) and human newborn dermal fibro-blast cells (green). As can be observed, the bioprinting method has evolved to a higher-level technology for manufacturing complicated vascular-shaped structures capable of carrying a large number of cells and ECM proteins; yet, it still confronts a difficult problem in reconnecting the vascular framework to the host circularity systems.

To address these issues, Lee et al. (2014) developed a first-of-its-kind bioprinting technology to establish a framework employing capillaries to massive perfused vas-cular channels. The huge channel was created via 3D bioprinting, and the angiogenic sprouting of endothelial cells from the margin of the massive conduits was accom-plished by the natural maturation process. Because these printed frameworks are often thin or hollow in structure, they are fed by a diffusion process from the host vasculature. Furthermore, Norotte et al. (2009) used scaffold-free bioprinting to cre-ate macrovascular tubular structures from multicellular cylinders. The double-lay-ered vascular tubes, which are analogous to vessels, were created using green and red bioinks. Kirillova et al. (2017) devised another vascularization technology, 4D bio-printing of self-folding tubes. Thin films of methacrylated alginate and hydroxyapa-tite were produced and cross-linked using green visible light. When submerged with water, phosphate-buffered saline (PBS), or cell culture conditions, the hydrogel sheets instantly self-folded into tubes. Four-dimensional bioprinting technologies may enable the future construction of reconfigurable tissue structures with adjustable functionality and reactivity.

1.5 CONCLUSION AND FUTURE OUTLOOK

As a consequence, we can now generate body tissue, organs, 3D models, scaffold, and other components leveraging 3D bioprinting technology. Such bioprinting techniques are classified into a variety of technologies that may be used based on the application. These methods have been used in tissue engineering to develop implants such as porous scaffolds, cellular structures, biosynthetic tissues, and organs. This approach may be able to restore massive tissues and organs. It has been shown that fabrication of scaffolds has been efficiently made utilizing bioprinting based on extrusion, and ink- and laser-based 3D printing have trailed owing to software and hardware restrictions.

Multicomponent bioinks for cell growth, proliferation, and differentiation have aided several vascular systems. Gelatin-based polymers have also allowed cells to interact, develop, replicate, and diversify. Despite the 3D bioprinting technique being powerful, it has some limits that must be addressed, and other research methodologies must be developed to enhance it. Four-dimensional printing, like 3D printing, adds a fourth dimension of time in this context. In response to environmental factors, including pressure, heat, air, moisture, and radiation, these 4D-printed items may alter shape or function over time. Rapid breakthroughs in 4D innovation may enable us to build highly innovative components with a wide variety of uses in the near future. Miniaturization and sterilization of bioprinters and associated equipment, as well as making these systems user-friendly for end-user doctors, will be secondary design objectives in the future.

REFERENCES

ASM International. (2003). *Overview of Biomaterials and Their Use in Medical Devices.* ASM International, 1–11.

Babbar, A., Jain, V., & Gupta, D. (2019c). Neurosurgical Bone Grinding. In *Biomanufacturing* (pp. 137–155). https://doi.org/10.1007/978-3-030-13951-3_7

Babbar, A., Jain, V., & Gupta, D. (2019d). Thermo-mechanical aspects and temperature measurement techniques of bone grinding. *Materials Today: Proceedings, 33*, 1458–1462. https://doi.org/10.1016/j.matpr.2020.01.497

Babbar, A., Jain, V., & Gupta, D. (2020c). Preliminary investigations of rotary ultrasonic neurosurgical bone grinding using Grey-Taguchi optimization methodology. *Grey Systems: Theory and Application, 10*(4), 479–493. https://doi.org/10.1108/gs-11-2019-0054

Babbar, A., Jain, V., Gupta, D., & Agrawal, D. (2021d). Finite element simulation and integration of CEM43 °C and Arrhenius models for ultrasonic-assisted skull bone grinding: A thermal dose model. *Medical Engineering and Physics, 90*, 9–22. https://doi.org/10.1016/j.medengphy.2021.01.008

Babbar, A., Jain, V., Gupta, D., & Agrawal, D. (2021e). Histological evaluation of thermal damage to osteocytes: A comparative study of conventional and ultrasonic-assisted bone grinding. *Medical Engineering and Physics, 90*, 1–8. https://doi.org/10.1016/j.medengphy.2021.01.009

Babbar, A., Jain, V., Gupta, D., Agrawal, D., Prakash, C., Singh, S., Wu, L. Y., Zheng, H. Y., Królczyk, G., & Bogdan-Chudy, M. (2021a). Experimental analysis of wear and multishape burr loading during neurosurgical bone grinding. *Journal of Materials Research and Technology, 12*, 15–28. https://doi.org/10.1016/j.jmrt.2021.02.060

Babbar, A., Jain, V., Gupta, D., Prakash, C., Singh, S., & Sharma, A. (2020d). 3D Bioprinting in Pharmaceuticals, Medicine, and Tissue Engineering Applications. In *Advanced Manufacturing and Processing Technology* (pp. 147–161). CRC Press. https://doi.org/10.1201/9780429298042-7

Babbar, A., Jain, V., Gupta, D., Prakash, C., Singh, S., & Sharma, A. (2020e). Effect of Process Parameters on Cutting Forces and Osteonecrosis for Orthopedic Bone Drilling Applications. In *Characterization, Testing, Measurement, and Metrology* (pp. 93–108). https://doi.org/10.1201/9780429298073-6

Babbar, A., Jain, V., Gupta, D., Singh, S., Prakash, C., & Pruncu, C. (2020a). Biomaterials and Fabrication Methods of Scaffolds for Tissue Engineering Applications. In *3D Printing in Biomedical Engineering* (pp. 167–186). Springer, Singapore. https://doi.org/10.1007/978-981-15-5424-7_8

Babbar, A., Rai, A., & Sharma, A. (2021). Latest trend in building construction: Three-dimensional printing. *Journal of Physics: Conference Series, 1950*(1). https://doi.org/10.1088/1742-6596/1950/1/012007

Babbar, A., Sharma, A., Bansal, S., Mago, J., & Toor, V. (2019a). Potential applications of three-dimensional printing for anatomical simulations and surgical planning. *Materials Today: Proceedings, 33*, 1558–1561. https://doi.org/10.1016/j.matpr.2020.04.123

Babbar, A., Sharma, A., & Chugh, M. (2020b). Application of flexible sintered magnetic abrasive brush for finishing of brass plate. *Optimization in Engineering Research, 01*(01), 36–47. https://doi.org/10.47406/oer.2020.1106

Babbar, A., Sharma, A., Jain, V., & Jain, A. K. (2019b). Rotary ultrasonic milling of C/SiC composites fabricated using chemical vapor infiltration and needling technique. *Materials Research Express, 6*(8). https://doi.org/10.1088/2053-1591/ab1bf7

Babbar, A., Sharma, A., Kumar, R., Pundir, P., & Dhiman, V. (2021b). Functionalized Biomaterials for 3D Printing: An Overview of the Literature. In *Additive Manufacturing with Functionalized Nanomaterials* (pp. 87–107). https://doi.org/10.1016/b978-0-12-823152-4.00005-3

Babbar, A., Sharma, A., & Singh, P. (2021c). Multi-objective optimization of magnetic abrasive finishing using grey relational analysis. *Materials Today: Proceedings*. https://doi.org/10.1016/j.matpr.2021.01.004

Baraiya, R., Babbar, A., Jain, V., & Gupta, D. (2020). In-situ simultaneous surface finishing using abrasive flow machining via novel fixture. *Journal of Manufacturing Processes, 50*, 266–278. https://doi.org/10.1016/j.jmapro.2019.12.051

Bose, S., Ke, D., Sahasrabudhe, H., & Bandyopadhyay, A. (2018). Additive manufacturing of biomaterials. *Progress in Materials Science, 93*, 45–111. https://doi.org/10.1016/j.pmatsci.2017.08.003

Cao, J., Xiao, Z., Jin, W., Chen, B., Meng, D., Ding, W., Han, S., Hou, X., Zhu, T., Yuan, B., Wang, J., Liang, W., & Dai, J. (2013). Induction of rat facial nerve regeneration by functional collagen scaffolds. *Biomaterials, 34*(4), 1302–1310. https://doi.org/10.1016/j.biomaterials.2012.10.031

Chakraborty, S., Das, P. P., & Kumar, V. (2017). A grey fuzzy logic approach for cotton fibre selection. *Journal of the Institution of Engineers (India): Series E, 98*(1). https://doi.org/10.1007/s40034-017-0099-7

Chakraborty, S., Das, P. P., & Kumar, V. (2018). Application of grey-fuzzy logic technique for parametric optimization of non-traditional machining processes. *Grey Systems: Theory and Application, 8*(1), 46–68. https://doi.org/10.1108/gs-08-2017-0028

Chakraborty, S., & Kumar, V. (2021). Development of an intelligent decision model for non-traditional machining processes. *Decision Making: Applications in Management and Engineering, 4*(1), 194–214. https://doi.org/10.31181/dmame2104194c

Chakraborty, S., Kumar, V., & Ramakrishnan, K. R. (2019). Selection of the all-time best world XI test cricket team using the TOPSIS method. *Decision Science Letters*, *8*(1), 95–108. https://doi.org/10.5267/j.dsl.2018.4.001

Dong, Q., Zhang, M., Zhou, X., Shao, Y., Li, J., Wang, L., Chu, C., Xue, F., Yao, Q., & Bai, J. (2021). 3D-printed mg-incorporated PCL-based scaffolds: A promising approach for bone healing. *Materials Science and Engineering C*, *129*. https://doi.org/10.1016/j.msec.2021.112372

Gao, C., Peng, S., Feng, P., & Shuai, C. (2017). Bone Biomaterials and Interactions with Stem Cells. In *Bone Research* (Vol. 5). https://doi.org/10.1038/boneres.2017.59

Guzzi, E. A., Bischof, R., Dranseikiene, D., Deshmukh, D. V., Wahlsten, A., Bovone, G., Bernhard, S., & Tibbitt, M. W. (2021). Hierarchical biomaterials via photopatterning-enhanced direct ink writing. *Biofabrication*, *13*(4). https://doi.org/10.1088/1758-5090/ac212f

Hernandez, J. L., Park, J., Yao, S., Blakney, A. K., Nguyen, H. V., Katz, B. H., Jensen, J. T., & Woodrow, K. A. (2021). Effect of tissue microenvironment on fibrous capsule formation to biomaterial-coated implants. *Biomaterials*, *273*. https://doi.org/10.1016/j.biomaterials.2021.120806

Herzog, D., Seyda, V., Wycisk, E., & Emmelmann, C. (2016). Additive manufacturing of metals. *Acta Materialia*, *117*, 371–392. https://doi.org/10.1016/j.actamat.2016.07.019

Horner, P. J., & Gage, F. H. (2000). Regenerating the damaged central nervous system. *Nature*, *407*(6807), 963–970. https://doi.org/10.1038/35039559

Hu, B., Guo, Y., Li, H., Liu, X., Fu, Y., & Ding, F. (2021). Recent advances in chitosan-based layer-by-layer biomaterials and their biomedical applications. *Carbohydrate Polymers*, *271*. https://doi.org/10.1016/j.carbpol.2021.118427

Huang, Y., Van Dessel, J., Martens, W., Lambrichts, I., Zhong, W. J., Ma, G. W., Lin, D., Liang, X., & Jacobs, R. (2015). Sensory innervation around immediately vs. delayed loaded implants: A pilot study. *International Journal of Oral Science*, *7*, 49–55. https://doi.org/10.1038/ijos.2014.53

Huh, J. T., Yoo, J. J., Atala, A., & Lee, S. J. (2020). Three-Dimensional Bioprinting for Tissue Engineering. In *Principles of Tissue Engineering* (pp. 1391–1415). https://doi.org/10.1016/b978-0-12-818422-6.00076-9

Jędzierowska, M., Binkowski, M., Koprowski, R., & Wróbel, Z. (2021). Evaluation of the effect of a PCL/NaNoSiO2 Implant on Bone Tissue Regeneration Using X-ray Micro-Computed Tomography. In *Advances in Intelligent Systems and Computing* (Vol. 1186, pp. 107–117). https://doi.org/10.1007/978-3-030-49666-1_9

Jin, S., Xia, X., Huang, J., Yuan, C., Zuo, Y., Li, Y., & Li, J. (2021). Recent advances in PLGA-based biomaterials for bone tissue regeneration. *Acta Biomaterialia*, *127*, 56–79. https://doi.org/10.1016/j.actbio.2021.03.067

Kadambi, P., Luniya, P., & Dhatrak, P. (2021). Current advancements in polymer/polymer matrix composites for dental implants: A systematic review. *Materials Today: Proceedings*, *46*, 740–745. https://doi.org/10.1016/j.matpr.2020.12.396

Karst, D., & Yang, Y. (2006). Molecular modeling study of the resistance of PLA to hydrolysis based on the blending of PLLA and PDLA. *Polymer*, *47*(13), 4845–4850. https://doi.org/10.1016/j.polymer.2006.05.002

Kirillova, A., Maxson, R., Stoychev, G., Gomillion, C. T., & Ionov, L. (2017). 4D biofabrication using shape-morphing hydrogels. *Advanced Materials*, *29*(46). https://doi.org/10.1002/adma.201703443

Kolesky, D. B., Truby, R. L., Gladman, A. S., Busbee, T. A., Homan, K. A., & Lewis, J. A. (2014). 3D bioprinting of vascularized, heterogeneous cell-laden tissue constructs. *Advanced Materials*, *26*(19), 3124–3130. https://doi.org/10.1002/adma.201305506

Kubyshkina, G., Zupančič, B., Štukelj, M., Grošelj, D., Marion, L., & Emri, I. (2011). Sterilization effect on structure, thermal and time-dependent properties of polyamides. *Conference Proceedings of the Society for Experimental Mechanics Series, 3,* 11–19. https://doi.org/10.1007/978-1-4614-0213-8_3

Kumar, V., & Chakraborty, S. (2022). *Analysis of the surface Roughness Characteristics of EDMed Components Using GRA Method.* In *Proceedings of the International Conference on Industrial and Manufacturing Systems (CIMS-2020)* (pp. 461–478). Springer, Cham. https://doi.org/10.1007/978-3-030-73495-4_32

Kumar, V., Das, P. P., & Chakraborty, S. (2020b). Grey-fuzzy method-based parametric analysis of abrasive water jet machining on GFRP composites. *Sādhanā, 45*(1), 106. https://doi.org/10.1007/s12046-020-01355-9

Kumar, V., Diyaley, S., & Chakraborty, S. (2020a). Teaching-learning-based parametric optimization of an electrical discharge machining process. *Facta Universitatis, Series: Mechanical Engineering, 18*(2), 281–300. https://doi.org/10.22190/FUME200218028K

Kumar, V., Kalita, K., Chatterjee, P., Zavadskas, E. K., & Chakraborty, S. (2021). A SWARA-CoCoSo-based approach for spray painting robot selection. *Informatica, 0*(0), 1–20. https://doi.org/10.15388/21-INFOR466

Labonnote, N., Rønnquist, A., Manum, B., & Rüther, P. (2016). Additive construction: State-of-the-art, challenges and opportunities. *Automation in Construction, 72,* 347–366). https://doi.org/10.1016/j.autcon.2016.08.026

Lee, V. K., Lanzi, A. M., Ngo, H., Yoo, S. S., Vincent, P. A., & Dai, G. (2014). Generation of multi-scale vascular network system within 3D hydrogel using 3D bio-printing technology. *Cellular and Molecular Bioengineering, 7*(3), 460–472. https://doi.org/10.1007/s12195-014-0340-0

Li, X., Liu, X., Huang, J., Fan, Y., & Cui, F. Z. (2011). Biomedical investigation of CNT based coatings. *Surface and Coatings Technology, 206*(4), 759–766. https://doi.org/10.1016/j.surfcoat.2011.02.063

Liu, Z., Liu, X., & Ramakrishna, S. (2021). Surface engineering of biomaterials in orthopedic and dental implants: Strategies to improve osteointegration, bacteriostatic and bactericidal activities. *Biotechnology Journal, 16*(7). https://doi.org/10.1002/biot.202000116

Miller, J. S., Stevens, K. R., Yang, M. T., Baker, B. M., Nguyen, D. H. T., Cohen, D. M., Toro, E., Chen, A. A., Galie, P. A., Yu, X., Chaturvedi, R., Bhatia, S. N., & Chen, C. S. (2012). Rapid casting of patterned vascular networks for perfusable engineered three-dimensional tissues. *Nature Materials, 11*(9), 768–774. https://doi.org/10.1038/nmat3357

Mondy, W. L., Cameron, D., Timmermans, J. P., De Clerck, N., Sasov, A., Casteleyn, C., & Piegl, L. A. (2009). Computer-aided design of microvasculature systems for use in vascular scaffold production. Biofabrication, *1*(3). https://doi.org/10.1088/1758-5082/1/3/035002

Murphy, S. V., & Atala, A. (2014). 3D bioprinting of tissues and organs. *Nature Biotechnology, 32*(8), 773–785). https://doi.org/10.1038/nbt.2958

Norotte, C., Marga, F. S., Niklason, L. E., & Forgacs, G. (2009). Scaffold-free vascular tissue engineering using bioprinting. *Biomaterials, 30*(30), 5910–5917. https://doi.org/10.1016/j.biomaterials.2009.06.034

Peltola, M. J., Vallittu, P. K., Vuorinen, V., Aho, A. A. J., Puntala, A., & Aitasalo, K. M. J. (2012). Novel composite implant in craniofacial bone reconstruction. *European Archives of Oto-Rhino-Laryngology, 269*(2), 623–628. https://doi.org/10.1007/s00405-011-1607-x

Pérez-Merino, P., Dorronsoro, C., Llorente, L., Durán, S., Jiménez-Alfaro, I., & Marcos, S. (2013). In vivo chromatic aberration in eyes implanted with intraocular lenses. Investigative Ophthalmology and Visual Science, *54*(4), 2654–2661. https://doi.org/10.1167/iovs.13-11912

Prakash, C., Kumar, V., Mistri, A., Uppal, A. S., Babbar, A., Pathri, B. P., Mago, J., Sharma, A., Singh, S., Wu, L. Y., & Zheng, H. Y. (2021). Investigation of functionally graded adherents on failure of socket joint of FRP composite tubes. *Materials, 14*(21), 6365. https://doi.org/10.3390/ma14216365

Scheib, J., & Höke, A. (2013). Advances in peripheral nerve regeneration. *Nature Reviews Neurology, 9*(12), 668–676. https://doi.org/10.1038/nrneurol.2013.227

Sharma, A., Babbar, A., Jain, V., & Gupta, D. (2018). Enhancement of surface roughness for brittle material during rotary ultrasonic machining. *MATEC Web of Conferences, 249*. https://doi.org/10.1051/matecconf/201824901006

Sharma, A., Grover, V., Babbar, A., & Rani, R. (2020). A Trending Nonconventional Hybrid Finishing/Machining Process. In *Non-Conventional Hybrid Machining Processes* (pp. 79–93). CRC Press. https://doi.org/10.1201/9780429029165-5

Sharma, A., Kumar, V., Babbar, A., Dhawan, V., Kotecha, K., & Prakash, C. (2021). Experimental investigation and optimization of electric discharge machining process parameters using grey-fuzzy-based hybrid techniques. *Materials, 14*(19), 5820. https://doi.org/10.3390/ma14195820

Singh, D., Babbar, A., Jain, V., Gupta, D., Saxena, S., & Dwibedi, V. (2019). Synthesis, characterization, and bioactivity investigation of biomimetic biodegradable PLA scaffold fabricated by fused filament fabrication process. *Journal of the Brazilian Society of Mechanical Sciences and Engineering, 41*(3). https://doi.org/10.1007/s40430-019-1625-y

Stratton, S., Shelke, N. B., Hoshino, K., Rudraiah, S., & Kumbar, S. G. (2016). Bioactive polymeric scaffolds for tissue engineering. *Bioactive Materials, 1*(2), 93–108). https://doi.org/10.1016/j.bioactmat.2016.11.001

Tan, F., Zhu, Y., Ma, Z., & Al-Rubeai, M. (2020). Recent advances in the implant-based drug delivery in otorhinolaryngology. *Acta Biomaterialia, 108*, 46–54. https://doi.org/10.1016/j.actbio.2020.04.012

Tappa, K., & Jammalamadaka, U. (2018). Novel biomaterials used in medical 3D printing techniques. *Journal of Functional Biomaterials, 9*(1). https://doi.org/10.3390/jfb9010017

Tayyaba, S., Ashraf, M. W., Ahmad, Z., Wang, N., Afzal, M. J., & Afzulpurkar, N. (2021). Article fabrication and analysis of polydimethylsiloxane (PDMS) microchannels for biomedical application. *Processes, 9*(1), 1–31. https://doi.org/10.3390/pr9010057

Teo, A. J. T., Mishra, A., Park, I., Kim, Y. J., Park, W. T., & Yoon, Y. J. (2016). Polymeric biomaterials for medical implants and devices. *ACS Biomaterials Science and Engineering, 2*(4), 454–472. https://doi.org/10.1021/acsbiomaterials.5b00429

Terrada, C., Julian, K., Cassoux, N., Prieur, A. M., Debre, M., Quartier, P., Lehoang, P., & Bodaghi, B. (2011). Cataract surgery with primary intraocular lens implantation in children with uveitis: Long-term outcomes. *Journal of Cataract and Refractive Surgery, 37*(11), 1977–1983. https://doi.org/10.1016/j.jcrs.2011.05.037

Wang, X., Jiang, M., Zhou, Z., Gou, J., & Hui, D. (2017). 3D printing of polymer matrix composites: A review and prospective. *Composites Part B: Engineering, 110*, 442–458. https://doi.org/10.1016/j.compositesb.2016.11.034

Yuan, Y., Ai, F., Zang, X., Zhuang, W., Shen, J., & Lin, S. (2004). Polyurethane vascular catheter surface grafted with zwitterionic sulfobetaine monomer activated by ozone. *Colloids and Surfaces B: Biointerfaces, 35*(1), 1–5. https://doi.org/10.1016/j.colsurfb.2004.01.005

Zdrahala, R. J., & Zdrahala, I. J. (1999). Biomedical applications of polyurethanes: A review of past promises, present realities, and a vibrant future. *Journal of Biomaterials Applications, 14*(1), 67–90. https://doi.org/10.1177/088532829901400104

Zhao, Y., Li, Y., Mao, S., Sun, W., & Yao, R. (2015). The influence of printing parameters on cell survival rate and printability in microextrusion-based 3D cell printing technology. *Biofabrication, 7*(4). https://doi.org/10.1088/1758-5090/7/4/045002

Zhou, J., Huang, X., Zheng, D., Li, H., Herrler, T., & Li, Q. (2014). Oriental nose elongation using an L-shaped polyethylene sheet implant for combined septal spreading and extension. *Aesthetic Plastic Surgery*, *38*(2), 295–302. https://doi.org/10.1007/s00266-014-0299-1

Zhu, Y., Gao, C., He, T., & Shen, J. (2004). Endothelium regeneration on luminal surface of polyurethane vascular scaffold modified with diamine and covalently grafted with gelatin. *Biomaterials*, *25*(3), 423–430. https://doi.org/10.1016/S0142-9612(03)00549-0

Zuback, J. S., & DebRoy, T. (2018). The hardness of additively manufactured alloys. *Materials*, *11*(11). https://doi.org/10.3390/ma11112070

2 Polymer 3D Bioprinting for Bionics and Tissue Engineering Applications

Vidyapati Kumar
Indian Institute of Technology, Kharagpur, India

Atul Babbar
Shree Guru Gobind Singh Tricentenary University,
Gurugram, India

Ankit Sharma
Chitkara University, Rajpura, India

Ranvijay Kumar
University Center for Research and Development,
Chandigarh University, Mohali, India

Ankit Tyagi
Shree Guru Gobind Singh Tricentenary University,
Gurugram, India

CONTENTS

DOI: 10.1201/9781003266464-2

2.1   INTRODUCTION

"3D Bioprinting" is a type of additive manufacturing that creates 3D structures that are functional 3D tissues using cells and biomaterials rather than typical metals and plastics. Bioinks are biomaterials that closely resemble the composition of our tissues. Bioprinting can be used in a variety of fields, including regenerative medicine, drug discovery and development, and 3D cell culture, to name a few. Organs-on-chips and other bioprinted constructs can be utilized to investigate human body functioning outside of the body in 3D. In comparison to an in vitro 2D model, the geometry of a 3D bioprinted structure is more akin to that of a naturally occurring biological system. Functional outcomes that are more physiologically relevant can be influenced by structural similarity. 3D bioprinting allows for a level of geometric intricacy in creating tissues that no other method allows for. That is why, by replacing animal testing and eliminating the organ transplant waiting list, this technology has the potential to dramatically transform the way we treat diseases. Bioprinting technologies can produce products that replicate the biological and functional qualities of naturally occurring structures and tissues in the human body. This could lead to a variety of applications, but for now, bioprinting has only one practical application: pharmaceutical drug testing and research. While the ultimate goal of 3D bioprinting is to create artificial organs for transplantation (as we'll see next), the difficulty of making them operate like genuine organs is enormous. Scientists can now successfully manufacture biological structures and tissues that are similar to those seen in nature. Researchers can now produce constructs that chemically behave like kidney tissue, rather than bioprinting entirely functional kidneys. While this is a long way from the original goal, these structures can be used to test new treatments without relying on real-life patients who may experience unanticipated adverse effects. Aside from the ethical considerations, medication research using bioprinted materials can reduce the expense of pre-clinical studies of new treatments, allowing them to be validated and commercialized sooner, while also potentially lowering the need for animal testing. 3D bioprinting could be a solution to global organ shortages and the growing aversion to testing new cosmetics, chemicals, and pharmaceuticals on animals. Although research efforts have increased quickly in recent years, it is unclear whether it will become a reality anytime soon. It goes without saying that printing organs is a "little bit" more difficult. Researchers discovered in the early 2000s that living cells may be sprayed through the nozzles of inkjet printers without causing damage. It is not enough to have cells; they also require a caring environment, which includes food, water, and oxygen. Microgels – think gelatin loaded with vitamins, proteins, and other life-sustaining substances – now provide these conditions.

Furthermore, researchers plant cells around 3D scaffolds comprising biodegradable polymers or collagen to create conditions that promote the fastest and most efficient cell growth, allowing them to grow into fully functional tissue.

Figure 2.1 represents a 3D bioprinting approach that starts with a medical image and finishes with printed tissue structures made with computer-aided design (CAD) or computer aided manufacturing (CAM) technology, as well as automated printing of 3D shapes that mimic target tissue or organs.

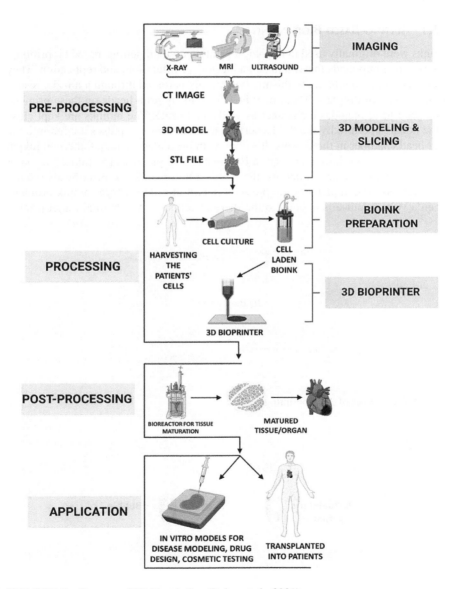

FIGURE 2.1 Process of 3D bioprinting (Bejoy et al., 2021).

2.2   VARIOUS POLYMER 3D BIOPRINTING APPROACHES

Several 3D bioprinting methods have been devised to bioengineer three-dimensional tissue or organ frameworks for biomedical applications, as depicted in Figure 2.2. The most widely employed varieties of 3D bioprinting techniques include extrusion-based bioprinting, laser-assisted bioprinting, and laser-based stereolithography. The effectiveness of each printing technique is heavily reliant on biomaterial choices and functions (Babbar, Jain, Gupta, Agrawal, et al., 2021a).

2.2.1   JETTING-BASED BIOPRINTING

Bioinks were originally used in the oldest printing method, jetting-based bioprinting. Bioinks are compounds that aid in cell adhesion, proliferation, and replication. They can be natural or synthetic. In this method, bioink is forced through a nozzle, resulting in a spray of droplets. There may be one or many print heads on these printers. Each print head includes a chamber as well as a nozzle. The bioinks are kept close to the nozzle aperture by the fluid's surface tension. Pressure pulses are fed into the print head chamber in three ways. It's done with piezoelectric inkjet, thermal inkjet, or electrostatic bioprinting, as seen in Figure 2.3. The piezoelectric inkjet's actuator provides pressure pulses to deposit the bioinks; however, some print heads require back pressure to complement the pressure pulses in order to form bioink droplets. When a voltage pulse is applied to the thermal actuator of a thermal inkjet printer,

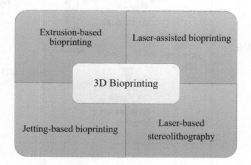

FIGURE 2.2   Types of 3D bioprinting.

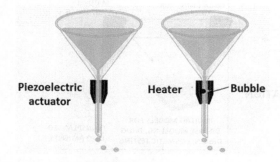

FIGURE 2.3   Jetting-based 3D bioprinting (Bejoy et al., 2021).

the bioink solution is heated locally. Local heating causes a vapour bubble, as seen in Figure 2.3 (Bejoy et al., 2021). The bioink droplet defies interfacial tension and accumulates on the scaffold as the bubble expands and contracts rapidly inside the fluid compartment, causing a force burst inside the fluid compartment. Biological materials such as proteins and mammalian cells, among other things, can be discharged using thermal inkjet printers. In electrostatic bioprinters, bioink droplets are formed by increasing the fluid compartment's capacity with the use of a bioink fluid attached to the plate. Once the voltage is applied, the pressure plate deflects between the electrode and the plate. Finally, the bioink is evacuated as the voltage declines and the pressure plate re-establishes its position, allowing printing to take place. Jetting-based bioprinting is a non-contact approach that uses picolitre bioink droplets deposited onto a substrate to create 2D and 3D structures. The process employed to form the bioink droplet can be classified as thermal, piezoelectric, laser-induced forward transfer, or pneumatic pressure in this sort of bioprinting. The thermal approach entails using a heat generator to raise temperature within the bioink chamber locally. A bubble forms as a result of the local heating, and a little droplet is ejected. A piezoelectric actuator is used in a bioprinter to apply a piezo-crystal pulse actuator mediated by electrical input, which results in the ejection of a tiny droplet. These two approaches are the most frequently utilized for jetting-based bioprinting, and they are used by marketed inkjet printers. The laser-induced forward transfer method, on the other hand, uses the laser system to cause vaporization, resulting in a small droplet. When compared to alternative delivery systems, this technology creates relatively high-resolution patterns; however, cell viability in the printed hydrogel is reduced. The droplet is created by the opening and shutting of a microvalve under constant pneumatic pressure in the bioprinter using pneumatic pressure. In comparison to other jetting-based bioprinting technologies, the principle of this technology is rather straightforward.

2.2.2 EXTRUSION-BASED BIOPRINTING

The idea behind extrusion-based bioprinting is to provide extrusion pressure to the bioink, which is good for tissue regeneration and repair. As shown in Figure 2.4, the bioink in this process is mostly deposited using pneumatic pressure, mechanical

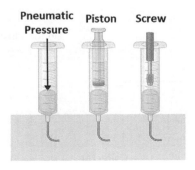

FIGURE 2.4   Extrusion-based 3D bioprinting (Bejoy et al., 2021).

pressure in the form of a screw or piston, and finally, the substrate is extruded out (Bejoy et al., 2021). The bioprinter's entire extrusion process is governed and controlled by the robotic stage controller. The bioink can be dispensed directly onto the substrate underneath the head in three directions: x, y, and z. High cell density bioinks, unique hydrogels with a wide range of viscosities, and biodegradable thermoplastics like polycaprolactone can all be dispensed by these printers. Extrusion bioprinting, as opposed to inkjet printers, lowers the risk of bioink clogging. The key disadvantage of extrusion is that we must ensure that the shear force isn't so high that cell viability is compromised.

2.2.3 LASER-ASSISTED BIOPRINTING

Another popular method for bioprinting living cells onto a substrate is laser-assisted bioprinting. A high-intensity light source or a light with a long wavelength is used to enable this printing. A laser bioprinter consists of a laser pulse, focusing lens, donor slide, energy absorption layer, donor substrate, and collector slide, as shown in Figure 2.5. The high-intensity light is concentrated by the focusing lens in Figure 2.5, after which the bioink is focused on the collection slide and the printed output is created. Laser printers, unlike inkjet printers, do not have nozzles and may thus deposit high densities of bioinks without clogging, as seen in Figure 2.5. (Bejoy et al., 2021).

2.2.4 LASER-BASED STEREOLITHOGRAPHY

As shown in Figure 2.6, it is a free-form procedure for depositing light onto cross-linked polymer materials (Bejoy et al., 2021). With an ultraviolet laser directed onto the surface of a liquid thermoset resin, the machine starts the 3D printing process by drawing the layers of the support structures, followed by the part itself. The build platform drops down once a layer is imaged on the resin surface, and a recoating bar moves across the platform to apply the next layer of resin. Layer by layer, the process is continued until the structure is complete. Newly constructed parts are removed from the machine and brought to a lab where any remaining resins are removed using solvents. The support structures are manually removed once the pieces are thoroughly clean. After that, pieces go through a UV-curing cycle to thoroughly firm

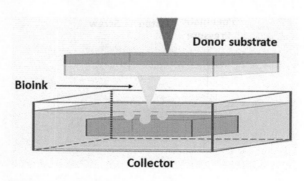

FIGURE 2.5   Laser-assisted bioprinting (Bejoy et al., 2021).

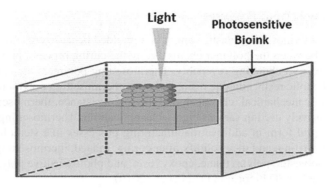

FIGURE 2.6   Laser-based stereolithography (Bejoy et al., 2021).

the part's exterior surface. The application of any custom or customer-specified finishing is the final stage in the SL process. To avoid degrading, parts made in SL should be utilized with limited UV and humidity exposure.

2.3   POLYMERS FOR 3D BIOPRINTING IN BIONICS AND TISSUE ENGINEERING

The properties of 3D bioprinting components are affected by both the material structure and the printing technique. Polymers are widely used as feed materials in 3D bioprinting owing to their simplicity of processing and the potential to produce cost-effective, functional products with tailored features and capabilities. Polymer selection for biomedical applications is influenced by a wide variety of mechanical qualities such as tensile strength, compressive strength, toughness, flexibility, biocompatibility, and biostability/biodegradability (Arefin et al., 2021). Thermoplastic, thermoset, elastomer, hydrogel, functional polymers, polymer blends, and polymer matrix composites may be employed in additive manufacturing to create biomedical components (Jiang et al., 2020).

2.3.1   THERMOPLASTIC POLYMERS

When heated, these materials soften and become moldable, and when cooled, they solidify to generate the desired shape. The procedure may be repeated because the long-chain polymers are held together by a weak Vander Waals force. Thermoplastic polymers include polyethylene, polystyrene, nylons, polycarbonate, and polyether imide, and thermoplastic products are manufactured employing conventional methods such as injection molding and material extrusion. However, leveraging thermoplastic polymer, 3D printing technologies such as fused deposition modelling (FDM) and selective laser sintering (SLS) can manufacture complex objects. Thermoplastic polymers often used in FDM include acrylonitrile butadiene styrene, polylactide, polyamide, polycarbonate, glass-filled nylon, polyether ester ketone (PEEK), and polyetherimide (Alghamdi et al., 2021). Polyamides and thermoplastic polyurethanes are often used in SLS and multi-jet techniques (Schmid & Wegener, 2016).

2.3.2 THERMOSETTING POLYMERS

Once molded, thermosetting plastic cannot be remolded because cross-linking links are generated between the polymer chains during the curing process, preventing the polymer chain from changing reversibly. Thermosetting polymers are epoxy, polyester resin, phenolic resin, silicone resin, and other thermosetting polymers. Because of their superior mechanical, chemical, and thermal resistance, thermosetting polymers are extensively used in various high-tech applications. Thermosetting polymers are used in liquid form in additive manufacturing processes like stereolithography and direct ink writing, and they solidify after curing. Indeed, thermosetting acrylate, cross-linked polyester, polyurethane, epoxy resin, and photosensitive polymer resins are often utilized in 3D bioprinting (Lei et al., 2019).

2.3.3 ELASTOMERS

Elastomers are polymeric materials with exceptional stretch ability owing to their low intermolecular forces (Özdemir, 2020). Natural rubber, synthetic polyisoprene, butyl rubber, epichlorohydrin rubber, silicone rubber, thermoplastic elastomers, and other elastomers are examples of elastomers. Silicon elastomer, thermoplastic elastomer, hydrogel elastomer, liquid crystal elastomer, polydimethylsiloxane elastomer, and polydimethylsiloxane elastomer are some of the elastomers utilized in 3D bioprinting (Babbar, Jain, Gupta, Agrawal, et al., 2021a; Babbar, Sharma, Kumar, et al., 2021b; Kumar et al., 2020; Singh et al., 2019; Zhou et al., 2019). Three-dimensional printed constructs with higher mechanical characteristics are needed for various load-bearing components. However, 3D-printed polymer products are weaker than traditionally made things. This issue may be solved by producing 3D-printed polymer composite or nanocomposite components (Babbar, Jain, Gupta, et al., 2020e; Hmeidat et al., 2020). Composite materials are made up of at least two distinct phases separated by a discrete interface. Different phases are correctly mixed to form a material system with superior structural or functional qualities that individual materials cannot accomplish. Different kinds of reinforcing materials may be used to improve the qualities of 3D-printed polymer composite products as depicted in Table 2.1. Polymer is the continuous phase of polymer matrix composites, whereas reinforcements are discontinuous. Reinforcement is physically and chemically connected with the matrix phase while retaining individuality. Mechanical characteristics of composite materials are determined by the properties of the matrix polymer, reinforcing elements, their concentration, and the interfacial contact between two distinct phases. Particle-reinforced polymer composites are made up of particle reinforcement in a polymer matrix. Particles vary in form, size, and morphology. The selection of reinforcing materials is determined by the predicted qualities of the composite (Wickramasinghe et al., 2020). Polymer serves as the matrix material in polymer nanocomposites, whereas nanoparticles serve as reinforcing material. The combination of nanotechnology and additive manufacturing enables material scientists to create 3D objects with custom characteristics and multi-functionality (Babbar, Jain, Gupta, & Agrawal, 2021c; Babbar, Jain, Gupta, et al., 2020d; Babbar et al. 2020d; Wu et al., 2020).

TABLE 2.1

Characteristics Improvement in Bioprinted Produced Components by Reinforcing Suitable Material in the Polymer Matrix

Matrix Material	Reinforcement Material	Characteristics Improvement
Nylon blend Akasheh & Aglan (2019)	Chopped carbon fiber	Tensile strength, fracture resistance
Epoxy resin Agarwal et al. (2018)	Fiber glass	Tensile strength, fatigue life
Nylon 11 Chung & Das (2008)	Silica	Tensile and compressive properties
Epoxy Sandoval & Wicker (2006)	Multi-walled carbon nanotube	Tensile strength, fracture resistance
ABS Wei et al. (2015)	Graphene	Thermal properties
Epoxy acrylate Xu et al. (2019)	TiO_2	Tensile properties, flexural, and hardness
ABS Tekinalp et al. (2014)	Carbon fiber	Tensile properties
Nylon O'Connor & Dowling (2019)	Carbon, glass, and Kevlar	Inter-laminar shear strength
ABS Compton et al. (2017)	Carbon fiber	Thermal property

ABS: Acrylonitrile-butadiene-styrene

2.3.4 FIBER REINFORCEMENTS

A fibrous polymer composite is reinforced with either continuous or chopped (whiskers) fibers in a matrix material. Whisker fibers are short and stubby, whereas continuous fibers are lengthy. Natural and synthetic fibers are the two types of fibers that may be used in fiber-reinforced polymer composites' bioprinting (Hofstätter et al., 2017). Carbon fiber, glass fiber, and Kevlar are examples of synthetic fibers that may enhance the mechanical qualities of final 3D-printed polymer products. Natural fibers come from either plants or animals. Cellulose fibers are often utilized as reinforcement (Hofstätter et al., 2017). The biodegradability of several biodegradable polymers utilized in manufacturing biomedical components is not harmed by natural fiber reinforcing. As a result, natural fiber reinforcing may be employed for biodegradable biomaterial objects to increase mechanical qualities while minimizing the impact on polymer biodegradability (Nath & Nilufar, 2020). Compared with unreinforced polymers, fiber-reinforced additive produced items have considerably better mechanical characteristics (Agarwal et al., 2018; Akasheh & Aglan, 2019).

2.3.5 PARTICLE REINFORCEMENTS

A particulate composite is made up of particles suspended in a matrix. Particulates are divided into two types: flake and filled/skeletal. Particle reinforcements are commonly used in the manufacturing of 3D-printed components owing to their inexpensive cost and ease of processing with polymeric materials. For example, ceramic particles such as alumina (Al_2O_3) may be used for different polymer matrices in the 3D printing of polymer composites (Babbar, Rai, & Sharma, 2021e).

2.3.6 Nanomaterial's Reinforcements

Nanocomposites are materials formed by incorporating nanoparticles (also known as nanoreinforcements) into a macroscopic sample material (often referred to as the matrix). Using nanoparticles as reinforcing materials has allowed for the enhancement of attributes such as mechanical, electrical, thermal, and chemical properties for 3D-bioprinted components (Nath & Nilufar, 2020). Carbon nanotubes, graphene, nanoclay, SiC, nanocellulose, and other nanoreinforcing materials may all be employed in 3D bioprinting (Wu et al., 2020). Table 2.2 highlights some polymers that may be printed using various bioprinting processes and are extensively used in biomedical applications due to their biomimetic functionality. The homogeneous dispersion and distribution of nanoparticles in the polymer matrix and their strong interfacial contact are critical for the uniform characteristics of composite materials. The benefits of utilizing nanoreinforcing materials in biomedical applications include increased biocompatibility, improved physical or chemical properties of the scaffold, and increased tissue development surrounding the implant (Babbar, Prakash, Singh, et al., 2020b; Babbar, Sharma, Bansal, et al., 2019a).

TABLE 2.2

Utilization of Various Polymers with Their Biomimetic Functionality in Biomedical Applications

Polymers	Biomimetic Functionality	Biomedical Applications	Bioprinting Techniques	Advantages
PEG-based fibroblasts Christensen et al. (2015)	Cylindrical frameworks	Vascular frameworks	Extrusion	Biocompatibility
PLA- or PGA-based scaffolds Ding et al. (2013)	Natural rigid bonelike architecture	Cartilage bone	3D-based printing	Biocompatibility, rigidity
PEG- and βTCP-based scaffolds Zhang et al. (2015)	Biologically infused interfacial designs	Cartilage bone	Stereolithography	Biocompatibility
PLA screw-like scaffold Liu et al. (2016)	Native bone	Bone	3D-based printing	Bioactive, mechanical features
Chitosan-based polymers Mukhopadhyay et al. (2014)	Shape memory function	Pharmaceutical distribution, bone regeneration therapeutics	4D-based printing	Biocompatibility; improved medication release control; tunable mechanical characteristics
PCL-based Gelatin Cui et al. (2021)	Cylindrical frameworks	Bilayers; cell-laden bioscaffolds for tissue engineering	4D-based printing	Sustainable and biocompatible
PEG Cui et al. (2021)	Shape memory function	Cell-laden bilayers	4D-based printing	Biocompatibility

PEG: Polyethylene glycol, PLA: Polyglycolic acid, PGA: Polylactic acid, βTCP: βtricalcium phosphate, PCL: Polycaprolactone.

2.4 APPLICATIONS OF POLYMERS IN BIONICS AND TISSUE ENGINEERING

Three-dimensional bioprinting is a fast-expanding process with immense potential in a wide range of sectors, including aerospace, engineering, art, education, and medicine. This is a versatile technique that can print a variety of dimensions and materials to create multipurpose 3D structure systems (Jiang et al., 2020). Because of its capacity to modify medical objects, cost efficiency, design flexibility, and other benefits, 3D-printed things are frequently employed in numerous biomedical applications (Wang et al., 2017). Biocompatibility and biodegradability/biostability are two crucial qualities for materials for biomedical applications. For 3D printing, biomaterial components, both biodegradable and biostable polymers, are employed. Biodegradable polymers are used in soft tissue engineering, in which the tissue develops at the same pace that the scaffolds decompose. For structural implants, however, biostable polymers are employed.

Three-dimensional bioprinting is showing great potential in biomedical applications, and several studies have been conducted to address the machining and grinding of neurosurgical bone in order to recover its quality characteristics, which aids in the subsequent 3D bioprinting process (Babbar, Jain, & Gupta, 2020h; Babbar, Jain, Gupta, et al., 2020c; Babbar, Jain, Gupta, Singh, et al., 2020a; Babbar, Sharma, Jain, et al., 2019b; Baraiya et al., 2020; Sharma, Kalsia, et al., 2021a; Singh et al., 2021). Many optimization algorithms and multicriteria decision-making approaches are also used in this regard to generate optimized parametric settings in order to obtain the best 3D-bioprinted components (Babbar et al., 2020; Babbar, Sharma, & Singh, 2021d; Chakraborty et al., 2017, 2019; Chakraborty & Kumar, 2021; Kumar et al., 2020; Kumar et al., 2020; Kumar et al., 2021). Tailor-made polymeric biomaterial products derived from tomographic scans may be utilized to 3D print contoured things for patients in situations requiring tailored treatments, medications, tools, and organ replacements. In the biomedical area, polymeric materials may be employed for hard and soft tissue engineering and implant biomaterials. Improved mechanical characteristics are needed for a number of critical biomedical applications, including load-bearing soft tissue replacement, artificial muscles, and medical devices (Cristache et al., 2018; Jiang et al., 2020; Wu et al., 2020).

The absence of appropriate vasculature into implantable frameworks significantly restricts bioengineered tissue constructs. Furthermore, highly vascularized tissue constructions are required for the proper functioning and survival of tissues. (Huang et al., 2015; Mondy et al., 2009). When a three-dimensional tissue framework containing cells is implanted, the successful transmit of mass necessitates an intact microvascular network for sustaining the metabolically active activities of the cells inside the framework. The ingrowth of the microvascular network into the implanted bioengineered tissue framework, on the other hand, happens in a timely way, which is the most significant accomplishment for therapeutic reasons (Murphy & Atala, 2014). As a result, some researchers have acquired an interest and sought to manufacture and build the vascularized tissue framework as shown in Figure 2.7 using 3D bioprinting methods, which is a potential approach for this sort of application (Huh et al., 2020).

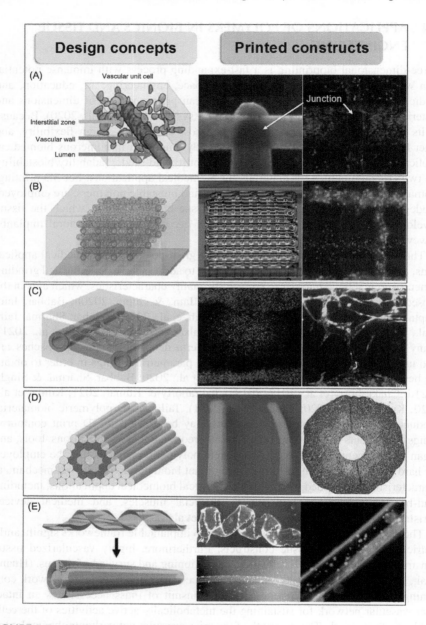

FIGURE 2.7   3D-bioprinted vasculature (Huh et al., 2020).

The simplest and most fundamental method of fabricating a vascular-shaped tissues framework is to use a sacrificial component that functions as a structure during printing and is then removed to leave a hollow-shaped tunnel-like structure. Miller et al. (2012) investigated a similar kind of work in which the vascular casting process was used using carbohydrate glass as a sacrificial template, as seen in Figure 2.7 (A).

The initial stage in this method is to inject a cell-based loaded hydrogel into a mold with a carbohydrate lattice structure, which is then removed once the hydrogel has been cross-linked. Human umbilical vein endothelial cells (HUVECs) are then injected into the lumen created to develop a vascular structure. Furthermore, extrusion-based bioprinting has been used successfully in this scenario to generate the vascular structure. Kolesky et al. (Kolesky et al., 2014) proposed an approach for fabricating a 3D framework with vasculature that includes many kinds of cells and ECM proteins, as shown in Figure 2.7 (B), using printing of two bioinks of HUVEC (red) and human newborn dermal fibroblast cells (green). As can be observed, the bioprinting method has evolved to a higher-level technology for manufacturing complicated vascular-shaped structures capable of carrying many cells and ECM proteins; yet, it still confronts a problematic problem in reconnecting the vascular framework to the host circularity systems.

Lee et al. [52] developed a first-of-its-kind bioprinting technology to establish a framework employing capillaries to massive perfused vascular channels to address these issues, as shown in Figure 2.7 (C). The huge channel was created via 3D bioprinting, and the natural maturation process accomplished the angiogenic sprouting of ECs from the margin of the massive conduits. Because these printed frameworks are often thin or hollow in structure, they are fed by a diffusion process from the host vasculature. Furthermore, as Figure 2.7 (D) illustrated, Norotte et al. (2009) used scaffold-free bioprinting to create macrovascular tubular structures from multicellular cylinders. Analogous to vessels, the double-layered vascular tubes were made using green and red bioinks. Kirillova et al. (2017) devised another vascularization technology, 4D bioprinting of self-folding conduits, shown in Figure 2.7. (E). Thin films of methacrylate alginate and hydroxyapatite (HA) were produced and cross-linked using visible green light. When submerged with water, phosphate-buffered saline, or under cell culture conditions, the hydrogel sheets instantly self-folded into tubes. Four-dimensional bioprinting technologies may enable the future construction of reconfigurable tissue structures with adjustable functionality and reactivity.

Bionics is also known as bioinspired architectural, which investigates and creates engineering equipment and contemporary technologies using biological processes and mechanisms inherent in nature. Figure 2.8 depicts some of the most popular 3D-bioprinted bionic organs (Bejoy et al., 2021). Figure 2.8 (A) shows scaffold-free grafting produced by multicellular spheroids (Bejoy et al., 2021). The longitudinal view of a bioprinted artificial trachea is demonstrated in Figure 2.8 (B) (Bejoy et al., 2021). Figure 2.8 (C) displays a vertical view of a five-layered artificial trachea (Bejoy et al., 2021). The 3D bioprinted collagen heart is observed in Figure 2.8 (D) (Bejoy et al., 2021). Figure 2.8 (E) exhibits a cross-sectional image of a 3D-bioprinted heart with the left and right ventricles revealed (Bejoy et al., 2021). Figure 2.8 (F) portrays a 3D-bioprinted distal lung component containing hydrogel (Bejoy et al., 2021).

2.4.1 SCAFFOLD

Due to musculoskeletal, cardiovascular, and connective tissue damage and the need for replacement, there is a high level of interest in the fabrication of soft tissue scaffolds. Because of the differences in form, size, and strength of diverse tissues,

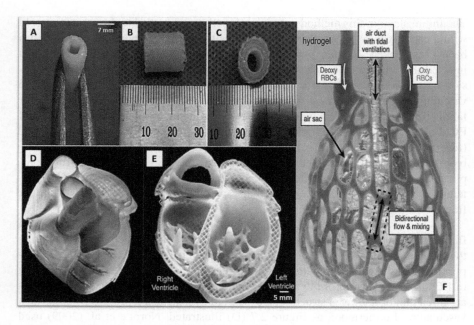

FIGURE 2.8   3D-bioprinted bionics (Bejoy et al., 2021).

accurately replacing these injured tissues is rather difficult. Tissue engineering scaffolds need tailoring of biological, mechanical, and chemical characteristics (Mondschein et al., 2017). Three-dimensional polymer printing is gaining popularity in tissue engineering scaffolds, where material qualities, design tactics, and procedures all play a role in designing the scaffold structure. Targeted tissue for 3D-printed materials may include bone, skin, ligaments, neurons, skeletal muscles, and so on (Arefin et al., 2021). For soft tissue replacement applications, the scaffold must deteriorate at a pace comparable to the regeneration of new tissues. Jackson et al. (2018) demonstrated a 3D-printed nylon-12 scaffold for cell differentiation, growth, and biomineral formation using SLS. Nylon-12 is biocompatible; however, SLS printing may provide a highly porous structure with a large surface area for surface modification and cell growth. Figure 2.9 (Liu et al., 2021) displays the animal surgery of rabbit in which the scaffold is inserted. Figures 9A and 9C show the abnormality in the rabbit knee joint, whereas Figures 9B and 9D exhibit the material implantation.

2.4.2   STENT

Stents are often utilized in percutaneous coronary intervention and subsequent nephrology operations. The huge stent is first folded to fit inside a catheter before being inserted into the blood artery. Shape memory polymers are utilized to manufacture polymer-based stents. The stent deploys within the blood artery using the polymer's shape memory effect. The shape memory material's super elasticity has

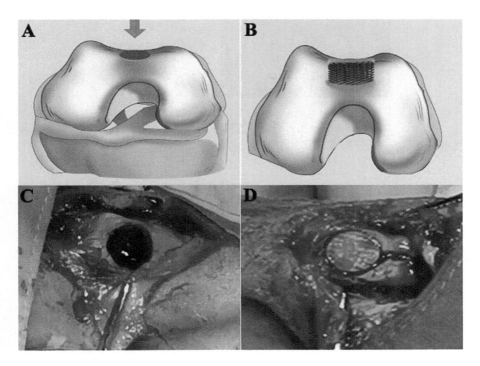

FIGURE 2.9   3D-printed scaffold inserted inside the rabbit for knee rehabilitation (Liu et al., 2021).

been used to produce a force that keeps the vessel open. Yeazel and Becker (2020) described 3D printing stents made from bioresorbable and shape-memory materials, which might remove the requirement for balloon expansion and minimize stent migration. Palliation therapy using an esophageal stent is the current standard of care for patients with inoperable esophageal cancer. Complications such as tumor ingrowth and stent migration into the stomach might result from using commonly used plastic stents for these purposes. Lin et al. (2019) described a mainly designed 3D-printed stent built with two polymers, such as flexible thermoplastic polyurethane, that offer flexibility for ease of deployment and self-expansion force to solve these difficulties. They chose poly(lactic acid) to optimize the mechanical qualities of the device so that it can open the clogged esophagus from the start. The poly(lactic acid) component may cause the placed stent to degrade slowly, eliminating the need for re-intervention (Lin et al., 2019). Figure 2.10 illustrates a 3D-printed prototype of a cardiovascular stent device developed using 3D bioprinting methods (Khalaj et al., 2021). It is reported that the stents were originally printed utilizing the material jetting approach in a variety of stent diameters, with the smallest stent fabricated being put on a balloon catheter for demonstration. The 3D-bioprinted prototype of a stent device for minimally invasive heart valve implant is then exhibited, and the stent structure was generated utilizing Projection Micro Stereolithography.

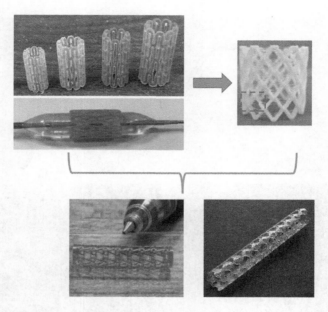

FIGURE 2.10   3D-printed prototype of a cardiovascular stent device (Khalaj et al., 2021).

2.4.3 PROSTHESES

Three-dimensional bioprinting of polymer may provide a broad range of prosthetics that are tailored to the individual's demands (Arefin et al., 2021). He et al. (2014) presented soft tissue prostheses such as an artificial nose, ear, and eye for maxillofacial rehabilitation utilizing a 3D bioprinting approach such as scanning, printing, polishing, and casting. Fabricating silicon prostheses is a straightforward and inexpensive method. Arjun et al. (2016) described 3D-printed hand prosthesis based on electrothermal actuators on nylon 6,6 muscles. A 3D-printed, manufactured prosthetic hand can move all fingers separately and grip various items. Zuniga developed antimicrobial 3D-printed finger prosthetics made of polylactic acid. Three-dimensional printing antimicrobial filament has the potential to change medical device creation. Stenvall et al. (2020) described a custom-made prosthetic device constructed from a polypropylene composite reinforced with micro-fibrillated cellulose. The prosthesis and orthosis solutions were created using the fused deposition modeling technology. They reported the clinical experiment regarding wearing experience, appearance, and material and method acceptability.

2.4.4 DENTISTRY

In the realm of dentistry, 3D-bioprinted manufactured components are becoming more widespread (Pillai et al., 2021). Vat polymerization method, digital light processing, polymer jetting, laser stereolithography, and fused deposition modeling are five additive manufacturing technologies typically used for 3D printing dental goods

FIGURE 2.11 Biocompatible 3D-printed PLA-based incisor teeth (Arun et al., 2020).

(Jockusch & Özcan, 2020). The process of vat polymerization may be used in implant dentistry (Nath & Nilufar, 2020). Because of their tunable qualities, vinyl polymers are often employed in dentistry. The majority are biocompatible and nondegradable, which are requirements for dental implants. Vinyl polymers are extensively utilized in sintering (e.g. SLS) or photo-polymerization in 3D printing of dental implants (e.g. SLA). Monomer, oligomer, and photo-initiator make up a photopolymer. The wavelength, strength, and period of radiation all affect the curing efficacy of photopolymers (Kim & Lee, 2020). Polyesters (such as polycaprolactone, polycarbonate, and polylactic acid) and polystyrene are also utilized in dental 3D printing (such as polystyrene and acrylonitrile-butadiene-styrene).

Polymer additive manufacturing for dental purposes may be done in various methods. Figure 2.11 shows the biocompatible 3D-printed PLA-based incisor teeth (Arun et al., 2020). The choice of material for bioprinted dental application is critical. Materials should be biocompatible, mechanically stable, and need little post-processing (Jockusch & Özcan, 2020). Other dental uses of 3D-printed polymers include guided tissue regeneration for periodontal abnormalities, prosthodontic crowns and bridges for provisional and permanent dental repair, fabrication of detachable prosthesis, and orthodontic micro screws.

2.5 CONCLUSION

Three-dimensional bioprinting is a fast-emerging field in medicine due to its inherent benefits, which include flexibility of use and a wide variety of applications in the biomedical field. Three-dimensional bioprinting processes provide design freedom, customization, and rapid prototyping. Furthermore, 3D bioprinting offers an excellent opportunity for designing complicated structures. This research provides an understanding of the 3D bioprinting of polymers for biomedical applications such as bionics and tissue engineering. A comprehensive overview of many 3D bioprinting procedures for polymers has been provided, as well as a list of important polymers and their 3D bioprinting characteristics. More study is needed, however, to improve the sluggish production of 3D bioprinting and the creation of novel functional polymers for biomedical applications and new processes to maintain additive manufacturing processes. Poor strength of 3D-printed components is a significant concern, and hence, the insertion of reinforcing materials technique may be used to improve the strength of the 3D-printed parts. Surface roughness, void development, and poor bond formation between reinforcement and polymer matrix are some of the issues that can arise during the fabrication of fiber-reinforced 3D-bioprinted components. Furthermore, additional research should be focused on creating 3D bioprinting components based on smart polymers that are nontoxic and exhibit the stimuli sensitive qualities necessary for biomedical applications. Although 3D bioprinting is a robust technology, it has significant limitations that must be addressed, and additional research approaches must be established to supplement it. In this context, 4D printing, like 3D printing, introduces a fourth dimension of time. These 4D-printed objects may change form or function as a result of external conditions such as pressure, heat, air, moisture, and radiation. Rapid advances in 4D innovation may allow us to construct highly inventive components with a broad range of applications in the near future.

REFERENCES

Agarwal, K., Kuchipudi, S. K., Girard, B., & Houser, M. (2018). Mechanical properties of fiber reinforced polymer composites: A comparative study of conventional and additive manufacturing methods. *Journal of Composite Materials*, 52(23), 3173–3181. https://doi.org/10.1177/0021998318762297

Akasheh, F., & Aglan, H. (2019). Fracture toughness enhancement of carbon fiber–reinforced polymer composites utilizing additive manufacturing fabrication. *Journal of Elastomers and Plastics*, 51(7–8), 698–711. https://doi.org/10.1177/0095244318817867

Alghamdi, S. S., John, S., Choudhury, N. R., & Dutta, N. K. (2021). Additive manufacturing of polymer materials: Progress, promise and challenges. *Polymers*, 13(5), 1–39). https://doi.org/10.3390/polym13050753

Arefin, A. M. E., Khatri, N. R., Kulkarni, N., & Egan, P. F. (2021). Polymer 3D printing review: Materials, process, and design strategies for medical applications. *Polymers*, 13(9). https://doi.org/10.3390/polym13091499

Arjun, A., Saharan, L., & Tadesse, Y. (2016). Design of a 3D printed hand prosthesis actuated by nylon 6-6 polymer based artificial muscles. *IEEE International Conference on Automation Science and Engineering, 2016-Novem*, 910–915. https://doi.org/10.1109/COASE.2016.7743499

Arun, M., Sathishkumar, N., Nithesh Kumar, K., Ajai, S. S., & Aswin, S. (2020). Development of patient specific bio-polymer incisor teeth by 3D printing process: A case study. *Materials Today: Proceedings*, 39, 1303–1308. https://doi.org/10.1016/j.matpr.2020.04.367

Babbar, A., Jain, V., & Gupta, D. (2020f). In vivo evaluation of machining forces, torque, and bone quality during skull bone grinding. *Proceedings of the Institution of Mechanical Engineers, Part H: Journal of Engineering in Medicine*, 234(6), 626–638. https://doi.org/10.1177/0954411920911499

Babbar, A., Jain, V., & Gupta, D. (2020h). Preliminary investigations of rotary ultrasonic neurosurgical bone grinding using Grey-Taguchi optimization methodology. *Grey Systems: Theory and Application*, 10(4), 479–493. https://doi.org/10.1108/gs-11-2019-0054

Babbar, A., Jain, V., Gupta, D., & Agrawal, D. (2021c). Finite element simulation and integration of CEM43 °C and Arrhenius Models for ultrasonic-assisted skull bone grinding: A thermal dose model. *Medical Engineering and Physics*, 90, 9–22. https://doi.org/10.1016/j.medengphy.2021.01.008

Babbar, A., Jain, V., Gupta, D., Agrawal, D., Prakash, C., Singh, S., Wu, L. Y., Zheng, H. Y., Królczyk, G., & Bogdan-Chudy, M. (2021a). Experimental analysis of wear and multi-shape burr loading during neurosurgical bone grinding. *Journal of Materials Research and Technology*, *12*, 15–28. https://doi.org/10.1016/j.jmrt.2021.02.060

Babbar, A., Jain, V., Gupta, D., Prakash, C., & Sharma, A. (2020c). Fabrication and Machining Methods of Composites for Aerospace Applications. In *Characterization, Testing, Measurement, and Metrology* (pp. 109–124). https://doi.org/10.1201/9780429298073-7

Babbar, A., Jain, V., Gupta, D., Prakash, C., Singh, S., & Sharma, A. (2020e). 3D Bioprinting in Pharmaceuticals, Medicine, and Tissue Engineering Applications. In *Advanced Manufacturing and Processing Technology* (pp. 147–161). https://doi.org/10.1201/9780429298042-7

Babbar, A., Jain, V., Gupta, D., Prakash, C., Singh, S., & Sharma, A. (2020g). Effect of Process Parameters on Cutting Forces and Osteonecrosis for Orthopedic Bone Drilling Applications. In *Characterization, Testing, Measurement, and Metrology* (pp. 93–108). https://doi.org/10.1201/9780429298073-6

Babbar, A., Jain, V., Gupta, D., & Sharma, A. (2020d). Fabrication of Microchannels Using Conventional and Hybrid Machining Processes. In *Non-Conventional Hybrid Machining Processes* (pp. 37–51). CRC Press. https://doi.org/10.1201/9780429029165-3

Babbar, A., Jain, V., Gupta, D., Singh, S., Prakash, C., & Pruncu, C. (2020a). Biomaterials and Fabrication Methods of Scaffolds for Tissue Engineering Applications. In *3D Printing in Biomedical Engineering* (pp. 167–186). Springer, Singapore. https://doi.org/10.1007/978-981-15-5424-7_8

Babbar, A., Prakash, C., Singh, S., Gupta, M. K., Mia, M., & Pruncu, C. I. (2020b). Application of hybrid nature-inspired algorithm: Single and bi-objective constrained optimization of magnetic abrasive finishing process parameters. *Journal of Materials Research and Technology*, *9*(4), 7961–7974. https://doi.org/10.1016/j.jmrt.2020.05.003

Babbar, A., Rai, A., & Sharma, A. (2021e). Latest trend in building construction: Three-dimensional printing. *Journal of Physics: Conference Series*, *1950*(1). https://doi.org/10.1088/1742-6596/1950/1/012007

Babbar, A., Sharma, A., Bansal, S., Mago, J., & Toor, V. (2019a). Potential applications of three-dimensional printing for anatomical simulations and surgical planning. *Materials Today: Proceedings*, *33*, 1558–1561. https://doi.org/10.1016/j.matpr.2020.04.123

Babbar, A., Sharma, A., Jain, V., & Jain, A. K. (2019b). Rotary ultrasonic milling of C/SiC composites fabricated using chemical vapor infiltration and needling technique. *Materials Research Express*, *6*(8). https://doi.org/10.1088/2053-1591/ab1bf7

Babbar, A., Sharma, A., Kumar, R., Pundir, P., & Dhiman, V. (2021b). Functionalized bio-materials for 3D printing: An overview of the literature. In *Additive Manufacturing with Functionalized Nanomaterials* (pp. 87–107). https://doi.org/10.1016/b978-0-12-823152-4.00005-3

Babbar, A., Sharma, A., & Singh, P. (2021d). Multi-objective optimization of magnetic abrasive finishing using grey relational analysis. *Materials Today: Proceedings*. https://doi.org/10.1016/j.matpr.2021.01.004

Baraiya, R., Babbar, A., Jain, V., & Gupta, D. (2020). In-situ simultaneous surface finishing using abrasive flow machining via novel fixture. *Journal of Manufacturing Processes*, *50*, 266–278. https://doi.org/10.1016/j.jmapro.2019.12.051

Bejoy, A. M., Makkithaya, K. N., Hunakunti, B. B., Hegde, A., Krishnamurthy, K., Sarkar, A., Lobo, C. F., Keshav, D. V. S., Dharshini, G., Dhivya Dharshini, S., Mascarenhas, S., Chakrabarti, S., Kalepu, S. R. R. D., Paul, B., & Mazumder, N. (2021). An insight on advances and applications of 3d bioprinting: A review. *Bioprinting*, *24*. https://doi.org/10.1016/j.bprint.2021.e00176

Chakraborty, S., Das, P. P., & Kumar, V. (2017). A grey fuzzy logic approach for cotton fibre selection. *Journal of the Institution of Engineers (India): Series E*, *98*(1). https://doi.org/10.1007/s40034-017-0099-7

Chakraborty, S., & Kumar, V. (2021). Development of an intelligent decision model for non-traditional machining processes. *Decision Making: Applications in Management and Engineering*, *4*(1), 194–214. https://doi.org/10.31181/dmame2104194c

Chakraborty, S., Kumar, V., & Ramakrishnan, K. R. (2019). Selection of the all-time best world XI test cricket team using the TOPSIS method. *Decision Science Letters*, *8*(1), 95–108. https://doi.org/10.5267/j.dsl.2018.4.001

Christensen, K., Xu, C., Chai, W., Zhang, Z., Fu, J., & Huang, Y. (2015). Freeform inkjet printing of cellular structures with bifurcations. *Biotechnology and Bioengineering*, *112*(5), 1047–1055. https://doi.org/10.1002/bit.25501

Chung, H., & Das, S. (2008). Functionally graded Nylon-11/silica nanocomposites produced by selective laser sintering. *Materials Science and Engineering A*, *487*(1–2), 251–257. https://doi.org/10.1016/j.msea.2007.10.082

Compton, B. G., Post, B. K., Duty, C. E., Love, L., & Kunc, V. (2017). Thermal analysis of additive manufacturing of large-scale thermoplastic polymer composites. *Additive Manufacturing*, *17*, 77–86. https://doi.org/10.1016/j.addma.2017.07.006

Cristache, C. M., Grosu, A. R., Cristache, G., Didilescu, A. C., & Totu, E. E. (2018). Additive manufacturing and synthetic polymers for bone reconstruction in the maxillofacial region. *Materiale Plastice*, *55*, 555–562. https://doi.org/10.37358/mp.18.4.5073

Cui, Y., Jin, R., Zhou, Y., Yu, M., Ling, Y., & Wang, L. Q. (2021). Crystallization enhanced thermal-sensitive hydrogels of PCL-PEG-PCL triblock copolymer for 3D printing. *Biomedical Materials (Bristol)*, *16*(3). https://doi.org/10.1088/1748-605X/abc38e

Ding, C., Qiao, Z., Jiang, W., Li, H., Wei, J., Zhou, G., & Dai, K. (2013). Regeneration of a goat femoral head using a tissue-specific, biphasic scaffold fabricated with CAD/CAM technology. *Biomaterials*, *34*(28), 6706–6716. https://doi.org/10.1016/j.biomaterials.2013.05.038

He, Y., Xue, G. H., & Fu, J. Z. (2014). Fabrication of low cost soft tissue prostheses with the desktop 3D printer. *Scientific Reports*, *4*. https://doi.org/10.1038/srep06973

Hmeidat, N. S., Pack, R. C., Talley, S. J., Moore, R. B., & Compton, B. G. (2020). Mechanical anisotropy in polymer composites produced by material extrusion additive manufacturing. *Additive Manufacturing*, *34*. https://doi.org/10.1016/j.addma.2020.101385

Hofstätter, T., Pedersen, D. B., Tosello, G., & Hansen, H. N. (2017). State-of-the-art of fiber-reinforced polymers in additive manufacturing technologies. *Journal of Reinforced Plastics and Composites*, *36*(15), 1061–1073). https://doi.org/10.1177/0731684417695648

Huang, Y., Van Dessel, J., Martens, W., Lambrichts, I., Zhong, W. J., Ma, G. W., Lin, D., Liang, X., & Jacobs, R. (2015). Sensory innervation around immediately vs. delayed loaded implants: A pilot study. *International Journal of Oral Science*, *7*, 49–55. https://doi.org/10.1038/ijos.2014.53

Huh, J. T., Yoo, J. J., Atala, A., & Lee, S. J. (2020). Three-dimensional bioprinting for tissue engineering. In *Principles of Tissue Engineering* (pp. 1391–1415). https://doi.org/10.1016/b978-0-12-818422-6.00076-9

Jackson, R. J., Patrick, P. S., Page, K., Powell, M. J., Lythgoe, M. F., Miodownik, M. A., Parkin, I. P., Carmalt, C. J., Kalber, T. L., & Bear, J. C. (2018). Chemically treated 3D printed polymer scaffolds for biomineral formation. *ACS Omega*, *3*(4), 4342–4351. https://doi.org/10.1021/acsomega.8b00219

Jiang, Z., Diggle, B., Tan, M. L., Viktorova, J., Bennett, C. W., & Connal, L. A. (2020). Extrusion 3D printing of polymeric materials with advanced properties. *Advanced Science*, *7*(17). https://doi.org/10.1002/advs.202001379

Jockusch, J., & Özcan, M. (2020). Additive manufacturing of dental polymers: An overview on processes, materials and applications. *Dental Materials Journal*, *39*(3), 345–354. https://doi.org/10.4012/dmj.2019-123

Khalaj, R., Tabriz, A. G., Okereke, M. I., & Douroumis, D. (2021). 3D printing advances in the development of stents. *International Journal of Pharmaceutics, 609.* https://doi. org/10.1016/j.ijpharm.2021.121153

Kim, J., & Lee, D. H. (2020). Influence of the postcuring process on dimensional accuracy and seating of 3D-printed polymeric fixed prostheses. *BioMed Research International, 2020.* https://doi.org/10.1155/2020/2150182

Kirillova, A., Maxson, R., Stoychev, G., Gomillion, C. T., & Ionov, L. (2017). 4D biofabrication using shape-morphing hydrogels. *Advanced Materials, 29*(46). https://doi. org/10.1002/adma.201703443

Kolesky, D. B., Truby, R. L., Gladman, A. S., Busbee, T. A., Homan, K. A., & Lewis, J. A. (2014). 3D bioprinting of vascularized, heterogeneous cell-laden tissue constructs. *Advanced Materials, 26*(19), 3124–3130. https://doi.org/10.1002/adma.201305506

Kumar, V., Das, P. P., & Chakraborty, S. (2020b). Grey-fuzzy method-based parametric analysis of abrasive water jet machining on GFRP composites. *Sādhanā, 45*(1), 106. https:// doi.org/10.1007/s12046-020-01355-9

Kumar, V., Diyaley, S., & Chakraborty, S. (2020a). Teaching-learning-based parametric optimization of an electrical discharge machining process. *Facta Universitatis, Series: Mechanical Engineering, 18*(2), 281–300. https://doi.org/10.22190/FUME200218028K

Kumar, V., Kalita, K., Chatterjee, P., Zavadskas, E. K., & Chakraborty, S. (2021). A SWARA-CoCoSo-based approach for spray painting robot selection. *Informatica, 0*(0), 1–20. https://doi.org/10.15388/21-INFOR466

Lei, D., Yang, Y., Liu, Z., Chen, S., Song, B., Shen, A., Yang, B., Li, S., Yuan, Z., Qi, Q., Sun, L., Guo, Y., Zuo, H., Huang, S., Yang, Q., Mo, X., He, C., Zhu, B., Jeffries, E. M., … You, Z. (2019). A general strategy of 3D printing thermosets for diverse applications. *Materials Horizons, 6*(2), 394–404. https://doi.org/10.1039/c8mh00937f

Lin, M., Firoozi, N., Tsai, C. T., Wallace, M. B., & Kang, Y. (2019). 3D-printed flexible polymer stents for potential applications in inoperable esophageal malignancies. *Acta Biomaterialia, 83,* 119–129. https://doi.org/10.1016/j.actbio.2018.10.035

Liu, A., Xue, G. H., Sun, M., Shao, H. F., Ma, C. Y., Gao, Q., Gou, Z. R., Yan, S. G., Liu, Y. M., & He, Y. (2016). 3D printing surgical implants at the clinic: A experimental study on anterior cruciate ligament reconstruction. *Scientific Reports, 6.* https://doi.org/10.1038/srep21704

Liu, J., Zou, Q., Wang, C., Lin, M., Li, Y., Zhang, R., & Li, Y. (2021). Electrospinning and 3D printed hybrid bi-layer scaffold for guided bone regeneration. *Materials and Design, 210.* https://doi.org/10.1016/j.matdes.2021.110047

Miller, J. S., Stevens, K. R., Yang, M. T., Baker, B. M., Nguyen, D. H. T., Cohen, D. M., Toro, E., Chen, A. A., Galie, P. A., Yu, X., Chaturvedi, R., Bhatia, S. N., & Chen, C. S. (2012). Rapid casting of patterned vascular networks for perfusable engineered three-dimensional tissues. *Nature Materials, 11*(9), 768–774. https://doi.org/10.1038/nmat3357

Mondschein, R. J., Kanitkar, A., Williams, C. B., Verbridge, S. S., & Long, T. E. (2017). Polymer structure-property requirements for stereolithographic 3D printing of soft tissue engineering scaffolds. *Biomaterials, 140,* 170–188). https://doi.org/10.1016/j. biomaterials.2017.06.005

Mondy, W. L., Cameron, D., Timmermans, J. P., De Clerck, N., Sasov, A., Casteleyn, C., & Piegl, L. A. (2009). Computer-aided design of microvasculature systems for use in vascular scaffold production. *Biofabrication, 1*(3). https://doi.org/10.1088/1758-5082/1/3/035002

Mukhopadhyay, P., Sarkar, K., Bhattacharya, S., Bhattacharyya, A., Mishra, R., & Kundu, P. P. (2014). PH sensitive N-succinyl chitosan grafted polyacrylamide hydrogel for oral insulin delivery. *Carbohydrate Polymers, 112,* 627–637. https://doi.org/10.1016/j. carbpol.2014.06.045

Murphy, S. V., & Atala, A. (2014). 3D bioprinting of tissues and organs. *Nature Biotechnology, 32*(8), 773–785). https://doi.org/10.1038/nbt.2958

Nath, S. D., & Nilufar, S. (2020). An overview of additive manufacturing of polymers and associated composites. *Polymers, 12*(11), 1–33). https://doi.org/10.3390/polym12112719

Norotte, C., Marga, F. S., Niklason, L. E., & Forgacs, G. (2009). Scaffold-free vascular tissue engineering using bioprinting. *Biomaterials*, *30*(30), 5910–5917. https://doi.org/10.1016/j.biomaterials.2009.06.034

O'Connor, H. J., & Dowling, D. P. (2019). Low-pressure additive manufacturing of continuous fiber-reinforced polymer composites. *Polymer Composites*, *40*(11), 4329–4339. https://doi.org/10.1002/pc.25294

Özdemir, T. (2020). Elastomeric micro- and nanocomposites for neutron shielding. In *Micro and Nanostructured Composite Materials for Neutron Shielding Applications* (pp. 125–137). https://doi.org/10.1016/b978-0-12-819459-1.00005-2

Pillai, S., Upadhyay, A., Khayambashi, P., Farooq, I., Sabri, H., Tarar, M., Lee, K. T., Harb, I., Zhou, S., Wang, Y., & Tran, S. D. (2021). Dental 3d-printing: Transferring art from the laboratories to the clinics. *Polymers*, *13*(1), 1–25. https://doi.org/10.3390/polym13010157

Sandoval, J. H., & Wicker, R. B. (2006). Functionalizing stereolithography resins: Effects of dispersed multi-walled carbon nanotubes on physical properties. *Rapid Prototyping Journal*, *12*(5), 292–303. https://doi.org/10.1108/13552540610707059

Schmid, M., & Wegener, K. (2016). Additive manufacturing: Polymers applicable for laser sintering (LS). *Procedia Engineering*, *149*, 457–464. https://doi.org/10.1016/j.proeng.2016.06.692

Sharma, A., Kalsia, M., Uppal, A. S., Babbar, A., & Dhawan, V. (2021a). Machining of hard and brittle materials: A comprehensive review. *Materials Today: Proceedings*. https://doi.org/10.1016/j.matpr.2021.07.452

Sharma, A., Kumar, V., Babbar, A., Dhawan, V., Kotecha, K., & Prakash, C. (2021b). Experimental investigation and optimization of electric discharge machining process parameters using grey-fuzzy-based hybrid techniques. *Materials*, *14*(19), 5820. https://doi.org/10.3390/ma14195820

Singh, D., Babbar, A., Jain, V., Gupta, D., Saxena, S., & Dwibedi, V. (2019). Synthesis, characterization, and bioactivity investigation of biomimetic biodegradable PLA scaffold fabricated by fused filament fabrication process. *Journal of the Brazilian Society of Mechanical Sciences and Engineering*, *41*(3). https://doi.org/10.1007/s40430-019-1625-y

Singh, G., Babbar, A., Jain, V., & Gupta, D. (2021). Comparative statement for diametric delamination in drilling of cortical bone with conventional and ultrasonic assisted drilling techniques. *Journal of Orthopaedics*, *25*, 53–58. https://doi.org/10.1016/j.jor.2021.03.017

Stenvall, E., Flodberg, G., Pettersson, H., Hellberg, K., Hermansson, L., Wallin, M., & Yang, L. (2020). Additive manufacturing of prostheses using forest-based composites. *Bioengineering*, *7*(3), 1–18. https://doi.org/10.3390/bioengineering7030103

Tekinalp, H. L., Kunc, V., Velez-Garcia, G. M., Duty, C. E., Love, L. J., Naskar, A. K., Blue, C. A., & Ozcan, S. (2014). Highly oriented carbon fiber-polymer composites via additive manufacturing. *Composites Science and Technology*, *105*, 144–150. https://doi.org/10.1016/j.compscitech.2014.10.009

Wang, X., Jiang, M., Zhou, Z., Gou, J., & Hui, D. (2017a). 3D printing of polymer matrix composites: A review and prospective. *Composites Part B: Engineering*, *110*, 442–458). https://doi.org/10.1016/j.compositesb.2016.11.034

Wang, Y. T., Yu, J. H., Lo, L. J., Hsu, P. H., & Lin, C. L. (2017b). Developing customized dental miniscrew surgical template from thermoplastic polymer material using image superimposition, CAD system, and 3D printing. *BioMed Research International*, *2017*. https://doi.org/10.1155/2017/1906197

Wei, X., Li, D., Jiang, W., Gu, Z., Wang, X., Zhang, Z., & Sun, Z. (2015). 3D printable graphene composite. *Scientific Reports*, *5*. https://doi.org/10.1038/srep11181

Wickramasinghe, S., Do, T., & Tran, P. (2020). FDM-based 3D printing of polymer and associated composite: A review on mechanical properties, defects and treatments. *Polymers, 12*(7), 1–42). https://doi.org/10.3390/polym12071529

Wu, H., Fahy, W. P., Kim, S., Kim, H., Zhao, N., Pilato, L., Kafi, A., Bateman, S., & Koo, J. H. (2020). Recent developments in polymers/polymer nanocomposites for additive manufacturing. *Progress in Materials Science, 111*. https://doi.org/10.1016/j.pmatsci.2020.100638

Xu, Y., Qi, S., & Xu, Y. (2019). Development of new TiO 2 nanoparticles modified photosensitive resin for rapid prototyping. *Science of Advanced Materials, 12*(2), 244–248. https://doi.org/10.1166/sam.2020.3604

Yeazel, T. R., & Becker, M. L. (2020). Advancing toward 3D printing of bioresorbable shape memory polymer stents. *Biomacromolecules, 21*(10), 3957–3965). https://doi.org/10.1021/acs.biomac.0c01082

Zhang, W., Lian, Q., Li, D., Wang, K., Hao, D., Bian, W., & Jin, Z. (2015). The effect of interface microstructure on interfacial shear strength for osteochondral scaffolds based on biomimetic design and 3D printing. *Materials Science and Engineering C, 46*, 10–15. https://doi.org/10.1016/j.msec.2014.09.042

Zhou, L. Y., Gao, Q., Fu, J. Z., Chen, Q. Y., Zhu, J. P., Sun, Y., & He, Y. (2019). Multimaterial 3D printing of highly stretchable silicone elastomers. *ACS Applied Materials and Interfaces, 11*(26), 23573–23583. https://doi.org/10.1021/acsami.9b04873

3 An Introduction to Bio-Implants and Biodegradable Materials
A Review

Tapinderjit Singh, Sandeep Singh, and Gurpreet Singh
Punjabi University, Patiala, India

CONTENTS

DOI: 10.1201/9781003266464-3

3.1 INTRODUCTION

In 1972, Mayan skulls were discovered by Amadeo Bobbio, some of them were more than 4000 years old. Nacre was substituted in place of teeth missing in the skull. Natural composite (Nacre) consists of 95–98 wt% of calcium carbonate with 2–5 wt% of organic matter. Biomedical (bioceramics) implantations began properly in the late 18th century for dental crowns using porcelain. In addition, gypsum (calcium sulfate dihydrate) was started for bone filling orthopedics in the late 19th century (Ruso et al., 2015). The major aim of biomedical materials is to match the appropriate combination of physical properties with replaced tissues, along with knowing the immune system of the host body. These implants enhanced the life of the individuals up to 25 years. Bioactive materials by the mid of 1980s reached at clinical use in a variety of applications such as orthopedic and dental along with various grouping of ceramics, bioglass, glass ceramics, and composites in powder and in coating forms (Hench and Polak, 2002).

Historically, gold is being used as biomaterial for dental repair based on literature: patient's teeth was repaired by hybrid animal teeth and linked by gold wire. Furthermore, properties of gold are chemically inert, malleable, ductile, resistant against chemical reactions, and nonsensitive to oxidation. Gold teeth prosthesis was elaborated by the end of the 19th century (Migonney, 2014). Since 1990, hydroxy-apatite (HA) composites are being used for repair and bone replacement. Bony apatite structure is closely related with HA as well as having similarity with the natural bone. Organic matrix is bounded with HA which plays a significant role in the bone's reconstruction. The patient can survive a lifetime with the prosthesis made from these biomaterials (Kattimani et al., 2016).

Materials such as alumina and zirconia are often considered as bioinert. The implant gets the shield of a soft tissue interlayer which makes no direct bone–material interface. A successful bone ingrowth by fibrous tissue is achieved only under compression with a porous structure to fit with bone cavity. Since 1990, use of zirconia femoral heads (6 million) and alumina components (3.5 million) worldwide led to implantation-associated clinical success (Bahraminasab et al., 2012). The major drawback reports associated with bioceramics are due its intrinsic brittleness. Alumina has been used with a limited number of designs under low mechanical loads

(Chevalier, 2006). The related mechanical properties of these materials are mentioned in Table 3.1 (Ruso et al., 2015). Moreover, yttria-stabilized zirconia (Y-TZP), due to high fracture toughness and strength, became in use having native structural similarity with alumina. Y-TZP enhanced new designs (knees and femoral heads) with improved mechanical properties which was previously not possible with alumina (Ruso et al., 2015). Figures 3.1–3.3 show the different sizes of femoral heads with the implant in the human body (Affatato, 2014b; Gautam and Malhotra, 2017; King and Phillips, 2016).

Millions of people suffer from bone diseases due to trauma, tumor, cancer, bone fractures (road accidents, soldiers injured during war, etc.), and miserably due to inadequate ideal bone replacement or treatment, and some of them are dying (Oakley, 2007). Therefore, research on biomaterials over the last four decades has expanded considerably for bone implantation, replacement, surgical reconstruction, and simulated prostheses to treat failure of organ or tissue. The biomaterial must incorporate with bone and possess biocompatibility, porosity, biomechanical compatibility, and osteoconductivity. Autografts and allografts are considered crucial for bone grafting method, providing osteoconductive and osteoinductive growth factors (Asgari and Cinvention, 2008).

TABLE 3.1
Mechanical Properties of Different Materials (Affatato 2014a)

Sr. no	Materials	Toughness (MPa)	Threshold (MPa)	Strength (MPa)	Vickers Hardness
1	Alumina	4.2	2.4	400–600	1800–2000
2	Zirconia	5.4	3.5	1000	1200–1300
3	Hydroxyapatite	0.9	0.6	50–60	500
4	Tricalcium phosphate	1.3	0.8	50–60	900
5	Micro–nano–Al_2–Zr	6	5	600	1800

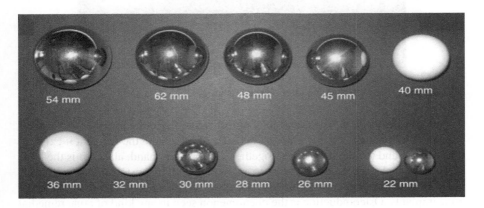

FIGURE 3.1   Different size of femoral heads of humans (Saverio Affatato 2014a).

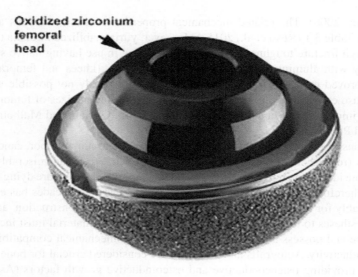

FIGURE 3.2   Oxidized zirconium femoral head (Kurtz and Ong 2016).

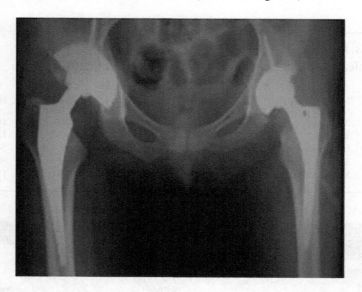

FIGURE 3.3   Femoral head hip joint (Cheung et al. 2015).

Autograft is the technique in which bone or tissue from one part is transferred to another on the patient's body. The success rate is high due to the fact that it exists in living tissue and their cells are kept together. On the other hand, allograft is the bone or tissue transplanted from one person to another. In addition, osteoconduction means the growth of new bone on the surface. The observable fact is usually seen in case of bone implants. Osteoinduction is the phenomenon seen in any kind of bone remedial process in case of fracture. This implies the use of undeveloped cells which develop into pre-osteoblast. Also, the healing rate of fracture depends on osteoinduction. However, autografts and autografting has their own disadvantages (Nacopoulos et al., 2014). Along with the advantages of autografting, there are also some limitations, which are as follows: (a) additional surgery, (b) problems in wound healing, (c) donor pain, (d) hazard of transmissible diseases like hepatitis and AIDS from tissues and fluids, and (e) tumor transplantation. The interest has created due to these limitations in the progress of synthetic materials in the applications of orthopedics and tissue regeneration. Hence, the materials used for these applications are classified as biomaterials.

3.2   BIOMATERIALS

Biomaterial is an engineered substance that interacts with a biological system for medical function either to repair, augment, and replace a natural function of a body, for example, HA-coated hip implants, teeth crowns in dental applications, heart valves, surgery, and drug delivery (Pal, 2014). In addition, biomaterials can be derived from nature or synthesized in laboratory using chemical approaches.

3.3   PROPERTIES OF BIOMATERIALS

(a) The material should not induce toxic or flammable response to implant in the body.
(b) Degradation time of the material should match with the curing or restoration process.
(c) The material should have suitable mechanical properties for indicated application.
(d) The degradation products should be nontoxic and able to be metabolized and cleared in the body.
(e) The material should have suitable permeability and processibility for projected purpose.

3.4   BIOMATERIALS ARE CLASSIFIED INTO FOUR CATEGORIES

(a) Metallic components
(b) Polymers
(c) Ceramics
(d) Composites

3.4.1   METALLIC COMPONENTS

Metallic biomaterials are engineered system scaffolds planned to bear natural tissues and are used commonly in joint replacements, dental implants, orthopedic fixations, neurosurgical devices, spine, cardiology, and stents. Most commonly used bioinert metals are Ti, Co, and stainless steel for load-bearing functions as well as resistance to corrosion and sustaining elongated period of steadiness and mechanical strength with negligible toxicity. These materials have excellent mechanical properties

(Prasad et al., 2017). The use of metallic implants maintains the same level of quality and activity of life in the host body. There are different metallic implants.

(a) Permanent metallic implants
(b) Tantalum-based bio-implants
(c) Stainless steel
(d) Titanium-based Ti alloys
(e) Magnesium alloys

3.4.1.1 Permanent Metallic Implants

The most frequently used metals for fracture fixation, angioplasty, and bone remodeling are surgical stainless steel (316L), cobalt-chromium (CoCr) alloys, and titanium (Ti), which are also categorized as bioinert metals. These metals have low corrosion, wear, and friction rate as well as long-term stability under highly reactive in vivo conditions. The osteolysis might destabilize the fixation and eventually the loading and force shift of the implant, leading to implant failure, corrective surgeries, or post-surgery complications (Yu et al., 1993). Figure 3.4 shows the permanent hip replacement implant in pelvis bone (Lawry et al., 2010).

3.4.1.2 Tantalum-Based Bioimplants

Tantalum is used under specialized conditions where high biocompatibility and corrosion resistance in acidic medium is needed. The stable natives of Ta_2O_5-protected film provide anticorrosion properties. Osseointegration and corrosion resistance are enhanced by coating stainless steel or titanium on artificial joints as porous tantalum has excellent bone-bonding properties (Cristea et al., 2015). Moreover, tantalum is used for load bearing orthopedic applications. However, the mechanical properties of bioactive materials such as HAs can be improved by coating on the same metal for orthopedic implants (Balla et al., 2010).

3.4.1.3 Stainless Steel 316L

Implants fabricated with pure metals often exhibit minor corrosion resistance and mechanical strength before the introduction of stainless steel (Muley et al., 2016). Powder metallurgy process is used for making parts of highly porous stainless steel 316L by selective laser sintering. Stainless steel 316L has biomedical properties like

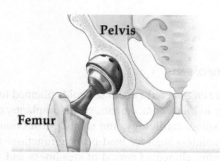

FIGURE 3.4   Hip replacement implant in pelvis bone (Gunatillake and Adhikari 2003).

biocompatibility, corrosion resistance, and high mechanical properties with sensible cost. Galvanic corrosion occurs due to the presence of Cl⁻ present in the blood plasma which corrodes stainless steel implants such as plates, nails, rods, etc. This corrosion provokes the discharge of Fe around the tissues and implants. Therefore, different coating techniques such as sol–gel are used to protect implants against corrosion and oxidation (Hosseinalipour et al., 2010).

3.4.1.4 Titanium Alloys

Titanium has properties such as biocompatibility, corrosion resistance, and lower modulus (Prakasam et al., 2017). Due to high specific strength, titanium alloys are used in dental applications as compared with other materials. Titanium β alloys are nontoxic along with higher strength and toughness. Hard tissue repair is mainly done by titanium alloys (Niinomi et al., 2012). Titanium alloys have 50% better strength-to-weight ratio compared to stainless steel and show better performance in medical grade and are suitable for higher loading applications. Weight plays an important role for the adjacent subjected bone. Bare Ti endorses cell integration, enabling much stronger bond between the implant and tissue in contrast to steel, which is due to the titanium dioxide layer formed on the surface with high dielectric constant. The strength of the Ti alloy can be increased by annealing, quenching, and thermal aging (Prasad et al., 2017).

3.4.1.5 Magnesium Alloys

Magnesium has the high strength-to-weight ratio and is the lightest metal with properties like biocompatibility, good shock absorption, and high damping capacity with good mechanical properties as compared with polymers. Unfortunately, magnesium is prone to corrosion but prevented by alloying with elements like aluminum and coatings (Hornberger et al., 2012). Magnesium alloys have biomedical applications in bone repair such as implanting screws, rods and metal plates, which, once implanted, would grant mechanical support and steadily degrade, providing room for increasing the bone tissue (Liu et al., 2018).

3.4.2 Polymers

Natural polymers and synthetic polymers are considered as biodegradable polymeric biomaterials (Gunatillake and Adhikari, 2003). Natural polymers are biomaterials used clinically and the rate of enzymatic degradation in vivo varies considerably with the spot of implantation depending upon the rate and availability of enzymes. The rate of degradation depends upon chemical modification (Nair and Laurencin, 2007). Natural polymers have inherent benefits like bioactivity, ability to present receptor-binding ligands to cells, and susceptibility to cell-triggered proteolytic degradation and natural remodeling.

Synthetic polymers are bioinert with predictable properties along with batch-to-batch reproducibility. Hydrolytically degradable polymers are favored due to negligible site-to-site and patient-to-patient variations in contrast to enzymatically degradable polymers. Biomaterial polymers are classified into hydrolytically and enzymatically degradable polymers (Maitz, 2015).

Hydrolytically degradable polymers are those chemical bonds which are hydrolytically broken down without the minor influence. The functional groups of hydrolysis include esters, orthoesters, anhydrides, carbonates, amides, etc. (Ulery et al., 2011).

3.4.3 Ceramics

Ceramic materials that are biocompatible are also categorized as bioceramics. Bioceramics vary in biocompatibility from ceramic oxides, which are inert in the body, to the supplementary excessive resorbable materials, which are ultimately swapped by body after they have assisted repair (Vallet-Regi and González-Calbet, 2004). Bioceramics are normally used in surgical implants, though bioceramics are flexible. Bioceramics are very much associated with body's own materials with enormously long-lasting metal oxides but not similar to porcelain.

Bioceramics are anticorrosive, biocompatible, and aesthetic in nature, which makes them suitable for medical usage. Zirconia is one of the bioceramic materials with bioinertness and noncytotoxicity. Their mechanical properties are similar to bone and they exhibit characteristics such as blood compatibility, no tissue reaction, and nontoxicity to cells, but bioinert ceramics do not make direct contact with bone (Stevens, 2008). On the other hand, the bioactivity of bioinert ceramics can be attained by forming composites. Calcium phosphate-reinforced ceramics are nontoxic to tissues, bioresorption, and increase osteoconduction. Moreover, ceramics have the benefit of being bioinert in the human body, and hardness and resistance to abrasion makes them helpful for teeth and bone replacements (Thamaraiselvi and Rajeswari, 2004).

3.4.3.1 Alumina

Alumina (Al_2O_3) is one of the ceramics that has a lifespan longer than the patient's lifespan. The material can be used in electrical insulation for pacemakers, catheter orifices, and in several prototypes of implantable systems such as cardiac pumps (Das and Bhattacharjee, 2019).

3.4.3.2 Calcium Phosphate

Calcium phosphate-based ceramic is used in orthopedic and maxillofacial surgery. The mineral phase composition is the same as that of the bone in structure. The material is suitable for bone implant due to the porosity of the material that enhances surface area for cell ionization. Figure 3.5 shows the permanent ceramic dental implant (Bisht et al., 2017 and Boch and Niepce, 2010).

3.4.3.3 Applications of Ceramics

(a) Dental and bone implants
(b) Kidney dialysis machine
(c) Pacemakers
(d) Bone regeneration

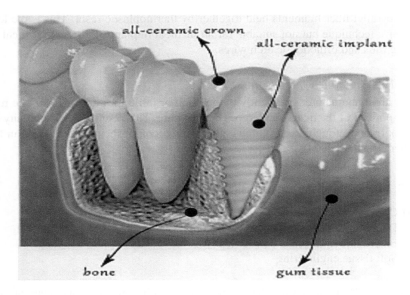

FIGURE 3.5 Permanent ceramic dental implant (Liu et al. 2018).

3.4.4 COMPOSITES

A composite material consists of two or more physically or chemically dissimilar, correctly assembled or distributed materials with an interface detaching them. It has uniqueness that is not described by any of the components in isolation; these specific characteristics being the intention of combining the materials. The continuous bulk phase is called the "matrix" and one or two dispersed noncontinuous phases are called reinforcement which has superior mechanical properties. The concept is to form a new material by combining two or more materials with performance and efficiency unattainable by individual constituents, and to enhance properties like strength, stiffness, toughness, and fatigue resistance of the material, for example, collagen fibers in an apatite (bone) (Fazeli et al., 2019). Due to this reason, biomedical composites are focused on the design of orthopedic and dental implants by weight savings, biocompatibility, property matching with natural structure, etc.

3.4.4.1 Types of Composites
  (a) Polymer matrix composites (PMC)
  (b) Ceramic matrix composites (CMC)

3.4.4.1.1 Polymer Matrix Composites
PMCs can be prepared by numerous ways, like manual layup of laminates, filament winding, and resin shift molding. Manual layup includes stacking pre-PEG tapes and

sheets parallel fiber filaments held together by thermoplastic resin. This is the least expensive technique but not suitable for all medical implants. Pultrusion is ideal for making rods and orthodontic arch wires.

3.4.4.1.2 Ceramic Matrix Composites

CMCs are manufactured frequently by pressing or infiltration techniques. The reinforced powder of the matrix undergoes hot pressing. This results in zero porosity but reduces fracture toughness. Alumina–glass dental composites are organized in this manner (Iftekhar, 2004).

3.4.5 BIOMEDICAL APPLICATIONS OF COMPOSITES

(a) Orthopedic
(b) Dental
(c) External prosthetics and orthotics
(d) Soft tissue engineering

Orthopedic: Bone fixation, plates, hip joints replacements, bone cement, and bone grafts are the orthopedic applications. Figure 3.6 shows the composites of Ti6A14V-HA for orthopedic implant (Miranda et al., 2016). Dental: Tooth crowns, repair of cavities, tooth replacement, etc. are dental applications. Soft tissue engineering: Crosslinked hydrogel, hydrogel poly(2-hydroxyethyl methacrylate), polylactide-co-glycolide, and poly glycolic acid are used in soft tissue engineering applications.

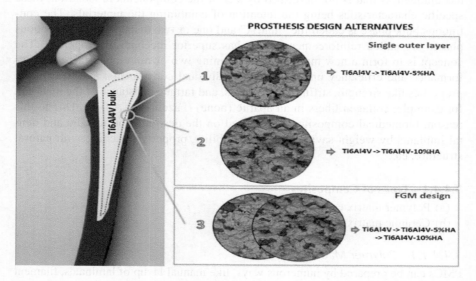

FIGURE 3.6 Demonstrative component design using Ti6A14V-HA composites (Mirzaei and Darroudi 2017).

3.5 TYPES OF BIOMATERIALS

The minority compounds are classified into bioactive, bioinert, and biodegradable; antibacterial and osteoconductive properties are listed below:

(a) Chitosan (CTS)
(b) Hydroxyapatite (HA)
(c) Bioglass
(d) Zirconium dioxide (ZrO_2)
(e) Magnesium (Mg)
(f) Titanium dioxide (TiO_2)
(g) Zinc oxide (ZnO)
(h) Polyvinylpyrrolidone (PVP)

3.5.1 CHITOSAN

CTS is one of the most rich natural polymers. CTS is derived from deacetylation of chitin (Islam et al., 2017). It is a natural polysaccharide found in crab, shrimp, lobster, coral, mushroom, fungi, etc. However, aquatic crustacean shells are broadly used as the principal sources for fabrication of CTS. Furthermore, broad segregation for CTS from marine crustaceans is completed by chemical hydrolysis method. Figure 3.7 shows the preparation of CTS from chitin by marine crustaceans (Younes and Rinaudo, 2015).

Steps involved in the chemical hydrolysis system to produce CTS are as follows:

(a) Demineralization
(b) Deproteinization
(c) Discoloration
(d) Deacetylation

CTS has a predominant contribution in biomedical applications due to its high bio-compatibility, biodegradability, binding ability, spongy configuration, antibacterial characteristic, nano-antigenicity, protein adsorption, etc. (Muxika et al., 2017). Although CTS has limitations, including oxidative stress and cytotoxicity, by adding grapheme oxide, these limitations can be eliminated (Ji et al., 2016). Also, CTS can be processed into membranes, sponges, scaffolds, nanoparticles, and nanofibers for biomedical applications. Figure 3.8 shows the structure of deacetylated CTS (Ahmed and Ikram, 2016).

3.5.1.1 Applications of Chitosan (Cheung et al., 2015 and Kanmani et al., 2017)

(a) Orthopedic applications
(b) Tissue engineering
(c) Wound healing
(d) New bone generation
(e) Anticancer treatment

```
┌──────────────────────┐
│   Marine Crustacean   │
└──────────────────────┘
          │
          ▼
┌──────────────────────┐
│     Deminerlization    │
└──────────────────────┘
          │
          ▼
┌──────────────────────┐
│    Deproteinization    │
└──────────────────────┘
          │
          ▼
┌──────────────────────┐
│      Discoloration     │
└──────────────────────┘
          │
          ▼
┌──────────────────────┐
│         Chitin         │
└──────────────────────┘
          │
          ▼
┌──────────────────────┐
│      Deacetylation     │
└──────────────────────┘
          │
          ▼
┌──────────────────────┐
│        Chitosan        │
└──────────────────────┘
```

FIGURE 3.7   Preparation of CTS by marine crustaceans (Mondal et al. 2018).

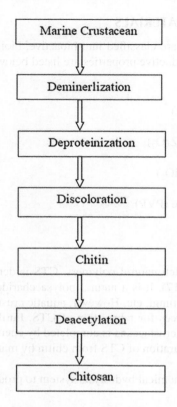

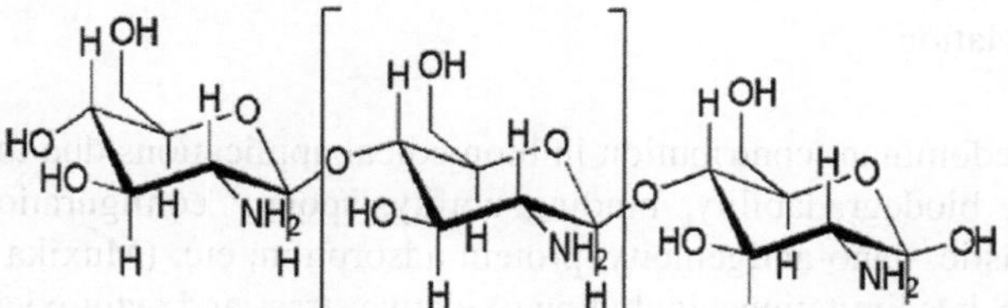

FIGURE 3.8   Structure of fully deacetylated CTS (Nacopoulos et al. 2014).

3.5.2 HYDROXYAPATITE

HA [$Ca_{10}(PO_4)_6(OH)_2$] is the most established form of calcium sulphate and is the mineral element of bone (60–65%). HA have similar morphological, crystallographic, size, and mechanical properties to that of human bone. It has also excellent properties such as biocompatibility, bioactivity, nontoxic, nonflammable, and osteoconductivity (Epple et al., 2010). It has been widely used in orthopedic, dental, coatings, and hard tissue repair. Natural HA is extracted from biological wastes such as natural bones (hen, goat, and cattle), egg shells, plants, and mineral sources (Pu'ad et al., 2019). However, it has low compressive strength, low density as well as low mechanical properties (Hidouri et al., 2019).

3.5.2.1 Applications of Hydroxyapatite

(a) Bone tissue engineering
(b) Bone void filler in orthopedic, traumatology, spine, maxillofacial, and dental surgery
(c) Orthopedic and dental implant coating
(d) Restoration of periodontal defects
(e) Drug and gene delivery

3.5.3 Bioglass

Bioglass also known as hench glass is poised of mol% 46.1 SiO_2, 24.2 Na_2O, 26.9 CaO, and 2.6 P_2O_5 is termed as bioglass (Lebecq et al., 2007). Bioglass bonds with living bone and forms an apatite layer on their surfaces after immersion in simulated body fluid. Due to its similar composition to HA (mineral component of bone), it provides bioglass the ability to be integrated with living bone. However, its low strength, brittleness, and nonload bearing applications raise the need to combine with materials (Ti, Zr, polylactic acid, etc.) that stabilize and enhance the mechanical properties without affecting the biocompatibility and bioactivity of bioglass (Hudecki et al., 2019).

3.5.3.1 Applications of Bioglass

(a) Bone regeneration
(b) Drug delivery
(c) Therapeutic molecule delivery
(d) Therapeutic ions doping

3.5.4 Zirconium Dioxide (ZrO_2)

ZrO_2 also known as zirconia is the white crystalline oxide of zirconium. It has high mechanical properties and used as a biomaterial for hipprothesis in orthopedic and dental implants (Singh et al., 2017). A new class of zirconia is formed by sintering that gains high mechanical and compressive strength than cortical bone. ZrO_2 has poor affinity toward cells and tissues, therefore the need of forming a composite requires with other ceramics like HA and bioglass. ZrO_2 has the property that it does not affect the biocompatibility of HA and bioglass but provides excellent strength and toughness to the composites (Mondal et al., 2018).

3.5.4.1 Applications of Zirconium Dioxide

(a) Tooth crowns and dental implants
(b) Bone restoration in orthopedics
(c) Manufacturing of ball heads for hip replacement

3.5.5 MAGNESIUM (MG)

Magnesium is a light weight metal with ionic presence and important functional roles in biological systems (Staiger et al., 2006). The density of magnesium is less than aluminum and steel. It is an essential element for human metabolism and is naturally found in bone tissue. Magnesium is a co-factor for enzymes and stabilizes DNA and RNA. In addition, magnesium has yield strength closer to natural bone, load bearing implants, osteoconductive and great fracture toughness than ceramic materials. Mg-based implants are biodegradable and get dissolved in the electrolytic physiological environment like body fluid after degradation that eliminates additional surgeries after bone healing (Yazdimamaghani et al., 2017). Unfortunately, magnesium has low corrosion resistance, but the rate of resistance is increased by using alloying elements and protective coatings (Agarwal et al., 2016).

3.5.5.1 Applications of Magnesium
  (a) Cardiovascular treatments
  (b) Tissue regeneration
  (c) Bone cell attachment
  (d) Bone regeneration

3.5.6 TITANIUM DIOXIDE (TiO$_2$)

Titanium dioxide is the naturally occurring oxide of titanium, also known as titania (Rahimi et al., 2016). Titanium and its alloys are used in biomedical engineering because of excellent corrosion resistance, biocompatibility, and high mechanical properties. Positive effects have been observed by functionalized TiO$_2$-based nano materials in biomedical applications; for instance, TiO$_2$ scaffolds increase the rate of apatite formation, biosensors, drug delivery, and osteoblast adhesion (Ribeiro et al., 2017).

Nanotubes of different diameter, length, and wall thickness of TiO$_2$ can release kinetic of specific drugs that are modified to accomplish steady and continuous release. TiO$_2$ nanotubes have good reproducibility and sensitivity to specific chemical and biochemical compounds, which is due to their high sensitivity to glucose, hydrogen peroxide, and cancer cells. Moreover, biosensors can sense blood glucose in patients with diabetes mellitus and enable early monitoring of cancer (Wu et al., 2014).

3.5.6.1 Applications of Titanium Dioxide
  (a) Intravascular stents
  (b) Bone tissue engineering
  (c) Biosensors
  (d) Drug delivery systems

3.5.7 ZINC OXIDE (ZnO)

Zinc oxide as a mineral of zincite exists under the earth crust and commercially formed by artificial methods (Mirzaei and Darroudi, 2017). ZnO nanoparticles have excellent compatibility to absorb ultra violet rays, are transparent to light, and are nontoxic. Zinc oxide plays a vital role in regeneration of the ROS cells after surpassing their antioxidative capacity. In addition, zinc oxide also has properties such as antimicrobial, anticancer activity, wound healing, biocompatibility, and insulin storage and secretion. $ZnSO_4$ and ZnO have antidiabetic activity and heal pancreatic injury (Mishra et al., 2017).

3.5.7.1 Applications of Zinc Oxide
(a) Anti-inflammatory
(b) Sunscreen agent
(c) Drug delivery
(d) Antidiabetic
(e) Anticancer

3.5.8 POLYVINYLPYRROLIDONE (PVP)

PVP is a water-soluble polymer made from the monomer of N-vinylpyrrolidone also known as polyvidone. It is an artificial polymer with superior biocompatibility and involves effortless synthesis procedure (Haaf et al., 1985). In addition, PVP has properties such as binding, nontoxicity, chemical inertness, pH-stability, film forming, and ability to form stable compounds (Kim et al., 2008). The structure of PVP is similar to proteins, enabling its use in biomedical operations, specifically in bone grafts and dental implants (Zhang et al., 2017).

3.5.8.1 Applications of Polyvinylpyrrolidone
(a) Wound healing
(b) Pain relief
(c) Binding agent
(d) Coating
(e) Solubility in alcohol, chloroform, methyl chloride, etc.

3.6 CONCLUSIONS

The present study presents a picture of ideal biomaterials and a research of important biomaterial features that are widely used for bioimplants. The biological, mechanical, and tribological qualities are the most important factors to consider when choosing biomaterials. On the other hand, low mechanical strength, low biostability, and poor corrosion resistance are the most common causes of incompetency in bioimplants. There are various types of biocompatible materials that have been used in recent years. The subsequent are the key conclusions drawn from this overview:

- HA is a bone grafting material that has good bonding strength and can be easily deposited on a substrate while preserving its bulk qualities.
- Bioactive glasses, which may arise in the future, are recapitulated here in terms of material bioinertness and bioactivity.

- The biological qualities of chitosan boost the healing processes of osteogenic progenitor cells with diverse growth factors and have tremendous potential for a wide range of applications.

To summarize, synergistic and multidisciplinary efforts are required to eliminate barriers associated with the merging of medical and technical sectors for the development of new-generation biomaterials.

REFERENCES

Affatato, S., 2014a. Contemporary Designs in Total Hip Arthroplasty (THA). In *Perspectives in Total Hip Arthroplasty* (pp. 46–64). Elsevier.

Affatato, S. ed., 2014b. *Perspectives in Total Hip Arthroplasty: Advances in Biomaterials and Their Tribological Interactions.* Woodhead Publishing, Elsevier, ISBN 978-1-78242-031-6.

Agarwal, S., Curtin, J., Duffy, B. and Jaiswal, S., 2016. Biodegradable magnesium alloys for orthopaedic applications: A review on corrosion, biocompatibility and surface modifications. *Materials Science and Engineering: C*, 68, pp. 948–963.

Ahmed, S. and Ikram, S., 2016. CTS based scaffolds and their applications in wound healing. *Achievements in the Life Sciences*, 10(1), pp. 27–37.

Asgari, S. and Cinvention A.G., 2008. Biodegradable therapeutic implant for bone or cartilage repair. U.S. Patent Application 12/098,722.

Bahraminasab, M., Sahari, B.B., Edwards, K.L., Farahmand, F., Arumugam, M. and Hong, T.S., 2012. Aseptic loosening of femoral components–a review of current and future trends in materials used. *Materials & Design*, 42, pp. 459–470.

Balla, V.K., Bose, S., Davies, N.M. and Bandyopadhyay, A., 2010. Tantalum—A bioactive metal for implants. *Jom*, 62(7), pp. 61–64.

Bisht, R., Mandal, A. and Mitra, A.K., 2017. Micro-and Nanotechnology-Based Implantable Devices and Bionics. In *Emerging Nanotechnologies for Diagnostics, Drug Delivery and Medical Devices* (pp. 249–290). Elsevier.

Boch, P. and Niepce, J.C. eds., 2010. *Ceramic Materials: Processes, Properties, and Applications* (Vol. 98). John Wiley & Sons.

Cheung, R., Ng, T., Wong, J. and Chan, W., 2015. CTS: an update on potential biomedical and pharmaceutical applications. *Marine Drugs*, 13(8), pp. 5156–5186.

Chevalier, J., 2006. What future for zirconia as a biomaterial? *Biomaterials*, 27(4), pp. 535–543.

Cristea, D., Ghiuta, I. and Munteanu, D., 2015. Tantalum based materials for implants and prostheses applications. *Bulletin of the Transilvania University of Brasov. Engineering Sciences. Series I*, 8(2), p. 151.

Das, R. and Bhattacharjee, C., 2019. Titanium-Based Nanocomposite Materials for Dental Implant Systems. In *Applications of Nanocomposite Materials in Dentistry* (pp. 271–284). Woodhead Publishing, Elsevier.

Epple, M., Ganesan, K., Heumann, R., Klesing, J., Kovtun, A., Neumann, S. and Sokolova, V., 2010. Application of calcium phosphate nanoparticles in biomedicine. *Journal of Materials Chemistry*, 20(1), pp. 18–23.

Fazeli, M., Florez, J.P. and Simão, R.A., 2019. Improvement in adhesion of cellulose fibers to the thermoplastic starch matrix by plasma treatment modification. *Composites Part B: Engineering*, 163, pp. 207–216.

Gautam, D. and Malhotra, R., 2017. Bilateral simultaneous total hip replacement in Achondroplasia. *Journal of Clinical Orthopaedics and Trauma*, 8, pp. S76–S79.

Gunatillake, P.A. and Adhikari, R., 2003. Biodegradable synthetic polymers for tissue engineering. *Eur Cell Mater*, 5(1), pp. 1–16.

Haaf, F., Sanner, A. and Straub, F., 1985. Polymers of N-vinylpyrrolidone: Synthesis, characterization and uses. *Polymer Journal*, 17(1), p. 143.

Hench, L.L. and Polak, J.M., 2002. Third-generation biomedical materials. *Science*, 295(5557), pp. 1014–1017.

Hidouri, M., Dorozhkin, S.V. and Albeladi, N., 2019. Thermal behavior, sintering and mechanical characterization of multiple ion-substituted HA bioceramics. *Journal of Inorganic and Organometallic Polymers and Materials*, 29(1), pp. 87–100.

Hornberger, H., Virtanen, S. and Boccaccini, A.R., 2012. Biomedical coatings on magnesium alloys–a review. *Actabiomaterialia*, 8(7), pp. 2442–2455.

Hosseinalipour, S.M., Ershad-Langroudi, A., Hayati, A.N. and Nabizade-Haghighi, A.M., 2010. Characterization of sol–gel coated 316L stainless steel for biomedical applications. *Progress in Organic Coatings*, 67(4), pp. 371–374.

Hudecki, A., Kiryczyński, G. and Łos, M.J., 2019. Biomaterials, Definition, Overview. In *Stem Cells and Biomaterials for Regenerative Medicine* (pp. 85–98). Academic Press.

Iftekhar, A., 2004. Biomedical Composites. *Standard Handbook of Biomedical Engineering and Design*. McGraw-Hill Companies.

Islam, S., Bhuiyan, M.R. and Islam, M.N., 2017. Chitin and CTS: Structure, properties and applications in biomedical engineering. *Journal of Polymers and the Environment*, 25(3), pp. 854–866.

Ji, H., Sun, H. and Qu, X., 2016. Antibacterial applications of graphene-based nanomaterials: Recent achievements and challenges. *Advanced Drug Delivery Reviews*, 105, pp. 176–189.

Kanmani, P., Aravind, J., Kamaraj, M., Sureshbabu, P. and Karthikeyan, S., 2017. Environmental applications of CTS and cellulosic biopolymers: A comprehensive outlook. *Bioresource Technology*, 242, pp. 295–303.

Kattimani, V.S., Kondaka, S. and Lingamaneni, K.P., 2016. HA—Past, present, and future in bone regeneration. *Bone and Tissue Regeneration Insights*, 7, pp. BTRI-S36138.

Kim, G.M., Asran, A.S., Michler, G.H., Simon, P. and Kim, J.S., 2008. Electrospun PVA/HA nanocomposite nanofibers: Biomimetics of mineralized hard tissues at a lower level of complexity. *Bioinspiration & Biomimetics*, 3(4), p. 046003.

King, A. and Phillips, J.R., 2016. Total hip and knee replacement surgery. *Surgery (Oxford)*, 34(9), pp. 468–474.

Kurtz, S.M., and Ong, K., 2016. Contemporary Total Hip Arthroplasty: Alternative Bearings. In *UHMWPE Biomaterials Handbook*, Jan 1 (pp. 72–105). William Andrew Publishing.

Lawry, G.V., Kreder, H.J., Hawker, G. and Jerome, D., 2010. *SPEC-Fam's Musculoskeletal Examination and Joint Injection Techniques Ebook (12-Month Access): Expert Consult-Online+ Print*. Elsevier Health Sciences.

Lebecq, I., Désanglois, F., Leriche, A. and Follet-Houttemane, C., 2007. Compositional dependence on the in vitro bioactivity of invert or conventional bioglasses in the Si-Ca-Na-P system. *Journal of Biomedical Materials Research Part A: An Official Journal of The Society for Biomaterials, The Japanese Society for Biomaterials, and The Australian Society for Biomaterials and the Korean Society for Biomaterials*, 83(1), pp. 156–168.

Liu, C., Ren, Z., Xu, Y., Pang, S., Zhao, X. and Zhao, Y., 2018. Biodegradable magnesium alloys developed as bone repair materials: A review. *Scanning*. doi: 10.1155/2018/9216314.

Maitz, M.F., 2015. Applications of synthetic polymers in clinical medicine. *Biosurface and Biotribology*, 1(3), pp. 161–176.

Migonney, V., 2014. History of biomaterials. *Biomaterials*, 3, pp. 1–10. ISBN:9781848215856

Miranda, G., Araújo, A., Bartolomeu, F., Buciumeanu, M., Carvalho, O., Souza, J.C.M., Silva, F.S. and Henriques, B., 2016. Design of Ti6Al4V-HA composites produced by hot pressing for biomedical applications. *Materials & Design*, 108, pp. 488–493.

Mirzaei, H. and Darroudi, M., 2017. Zinc oxide nanoparticles: Biological synthesis and biomedical applications. *Ceramics International*, 43(1), pp. 907–914.

Mishra, P.K., Mishra, H., Ekielski, A., Talegaonkar, S. and Vaidya, B., 2017. Zinc oxide nanoparticles: a promising nanomaterial for biomedical applications. *Drug Discovery Today*, 22(12), pp. 1825–1834.

Mondal, S., Hoang, G., Manivasagan, P., Moorthy, M.S., Nguyen, T.P., Phan, T.T.V., Kim, H.H., Kim, M.H., Nam, S.Y. and Oh, J., 2018. Nano-HA bioactive glass composite scaffold with enhanced mechanical and biological performance for tissue engineering application. *Ceramics International*, 44(13), pp. 15735–15746.

Muley, S.V., Vidvans, A.N., Chaudhari, G.P. and Udainiya, S., 2016. An assessment of ultra fine grained 316L stainless steel for implant applications. *Acta Biomaterialia*, 30, pp. 408–419.

Muxika, A., Etxabide, A., Uranga, J., Guerrero, P. and De La Caba, K., 2017. CTS as a bioactive polymer: Processing, properties and applications. *International Journal of Biological Macromolecules*, 105, pp. 1358–1368.

Nacopoulos, C., Dontas, I., Lelovas, P., Galanos, A., Vesalas, A.M., Raptou, P., Mastoris, M., Chronopoulos, E. and Papaioannou, N., 2014. Enhancement of bone regeneration with the combination of platelet-rich fibrin and synthetic graft. *Journal of Craniofacial Surgery*, 25(6), pp. 2164–2168.

Nair, L.S. and Laurencin, C.T., 2007. Biodegradable polymers as biomaterials. *Progress in Polymer Science*, 32(8–9), pp. 762–798.

Niinomi, M., Nakai, M. and Hieda, J., 2012. Development of new metallic alloys for biomedical applications. *Actabiomaterialia*, 8(11), pp. 3888–3903.

Oakley, A., 2007. *Fracture: Adventures of a Broken Body*. Policy Press.

Pal, S., 2014. *Design of Artificial Human Joints & Organs* (p. 23). Boston, MA: Springer US.

Prakasam, M., Locs, J., Salma-Ancane, K., Loca, D., Largeteau, A. and Berzina-Cimdina, L., 2017. Biodegradable materials and metallic implants—A review. *Journal of functional biomaterials*, 8(4), p. 44.

Prasad, K., Bazaka, O., Chua, M., Rochford, M., Fedrick, L., Spoor, J., Symes, R., Tieppo, M., Collins, C., Cao, A. and Markwell, D., 2017. Metallic biomaterials: Current challenges and opportunities. *Materials*, 10(8), p. 884.

Pu'ad, N.M., Koshy, P., Abdullah, H.Z., Idris, M.I. and Lee, T.C., 2019. Syntheses of HA from natural sources. *Heliyon*, 5(5), p. e01588.

Rahimi, N., Pax, R.A. and Gray, E.M., 2016. Review of functional titanium oxides. I: TiO2 and its modifications. *Progress in Solid State Chemistry*, 44(3), pp. 86–105.

Ribeiro, A.R., Gemini-Piperni, S., Alves, S.A., Granjeiro, J.M. and Rocha, L.A., 2017. Titanium Dioxide Nanoparticles and Nanotubular Surfaces: Potential Applications in Nanomedicine. In *Metal Nanoparticles in Pharma* (pp. 101–121). Springer, Cham.

Ruso, M.J., Sartuqui, J. and Messina, P.V., 2015. Multiscale inorganic hierarchically materials: towards an improved orthopaedic regenerative medicine. *Current Topics in Medicinal Chemistry*, 15(22), pp. 2290–2305.

Singh, S., Ramakrishna, S. and Singh, R., 2017. Material issues in additive manufacturing: A review. *Journal of Manufacturing Processes*, 25, pp. 185–200.

Staiger, M.P., Pietak, A.M., Huadmai, J. and Dias, G., 2006. Magnesium and its alloys as orthopedic biomaterials: a review. *Biomaterials*, 27(9), pp. 1728–1734.

Stevens, M.M., 2008. Biomaterials for bone tissue engineering. *Materials Today*, 11(5), pp. 18–25.

Thamaraiselvi, T. and Rajeswari, S., 2004. Biological evaluation of bioceramic materials-a review. *Carbon*, 24(31), p. 172.

Ulery, B.D., Nair, L.S. and Laurencin, C.T., 2011. Biomedical applications of biodegradable polymers. *Journal of Polymer Science Part B: Polymer Physics*, 49(12), pp. 832–864.

Vallet-Regi, M. and González-Calbet, J.M., 2004. Calcium phosphates as substitution of bone tissues. *Progress in Solid State Chemistry*, 32(1–2), pp. 1–31.

Wu, S., Weng, Z., Liu, X., Yeung, K.W.K. and Chu, P.K., 2014. Functionalized TiO2 based nanomaterials for biomedical applications. *Advanced Functional Materials*, 24(35), pp. 5464–5481.

Yazdimamaghani, M., Razavi, M., Vashaee, D., Moharamzadeh, K., Boccaccini, A.R. and Tayebi, L., 2017. Porous magnesium-based scaffolds for tissue engineering. *Materials Science and Engineering: C*, 71, pp. 1253–1266.

Younes, I. and Rinaudo, M., 2015. Chitin and CTS preparation from marine sources. Structure, properties and applications. *Marine Drugs*, 13(3), pp. 1133–1174.

Yu, J., Zhao, Z.J. and Li, L.X., 1993. Corrosion fatigue resistances of surgical implant stainless steels and titanium alloy. *Corrosion Science*, 35(1–4), pp. 587–597.

Zhang, P., Zhi, Y., Fang, H., Wu, Z., Chen, T., Jiang, J. and Chen, S., 2017. Effects of PVP-iodine on tendon-bone healing in a rabbit extra-articular model. *Experimental and Therapeutic Medicine*, 13(6), pp. 2751–2756.

4 Biocompatible and Bioactive Ceramics for Biomedical Applications
Content Analysis

Raman Kumar and Swapandeep Kaur
Guru Nanak Dev Engineering College, Ludhiana,
Punjab, India

CONTENTS

4.1 INTRODUCTION

A biomaterial is a material that interacts with biological structures to diagnose, treat, augment, or replace a bodily muscle, structure, or functionality. Therefore, the characteristics and biocompatibility of biomaterials are significant issues that must be addressed and overcome before introducing a biomaterial to the market or embedding it into a biological system. To improve the biocompatibility of biomaterials, numerous surface treatment procedures have been investigated, including physical and chemical, mechanical, and biologically alterations (John et al., 2015). Bioceramics and bioglasses are biocompatible ceramic materials. Bioceramics are a type and subdivision of biomaterials. Bioceramics are biocompatible in various ways, ranging from inert ceramic oxides to resorbable materials that are expelled by the body after assisting with a repair. As a result, bioceramics are used in many medical applications (Hench, 1993).

DOI: 10.1201/9781003266464-4

Biocompatibility refers to a biomaterial's capacity to behave in a variety of situations. The term refers to a material's capacity to function in a system with an appropriate host immunological response. The term's ambiguity reflects the continual advancement of knowledge about how biomaterials communicate with the human body and, consequently, how these actions impact a medical device's therapeutic outcome (Williams, 2008). For example, bioactive ceramics have been intended to stimulate particular bioactivity to restore afflicted organs. Bioactivity is understood as the power to make noticeable touch with living bone following implantation in bony defects to mend bone tissues. Although some bioceramics are versatile, they are commonly used as stiff materials in surgical implants. In addition, the ceramic materials used are not exclusive to those used in porcelain (Doremus, 1992).

On the other hand, bioceramics are either strongly connected to the body's possessed materials or are nigh-indestructible metal oxides (Eliaz & Metoki, 2017). Ceramic dental and bone implants are becoming more common in the medical field. Surgical cermets, comprised of ceramic and sintered metal, are commonly used. Bioceramic coatings are routinely used on joint replacements to minimise wear and inflammation (Guarino, Iafisco, & Spriano, 2020). Bioceramics are also used in pacemakers, kidney dialysis machines, and respirators, among other medical applications (Kinnari et al., 2009). In addition, bioceramics are used in endovascular circulation systems such as dialyzers or designed bioreactors. Due to their physicochemical properties, ceramics have many uses as biomaterials (Park, 2009). They are unique in a way that they are inactive in the human body, and their toughness and erosion tolerance make them ideal for bone and tooth replacement. In addition, many ceramics have high friction resistance, making them appealing as joint replacement materials (Fabbri, Celotti, & Ravaglioli, 1995). Properties like aesthetics and electrical insulation are also essential factors relating to bioceramics (John et al., 2015).

Alumina (Al_2O_3) is often used in bioceramics because it lasts longer than the patients' life. Middle ear ossicles, ocular prostheses, electrical insulation for implantable devices, catheters holes, and countless prototypes of implanted technologies such as cardiac pumps can all benefit from these materials (Semlitsch, Lehmann, Weber, Doerre, & Willert, 1977). Furthermore, whether pure or in ceramic–polymer composites, aluminosilicates are extensively used in dental prostheses. The ceramic–polymer composites could fill cavities instead of amalgams, which are harmful. Aluminosilicates have a glassy texture as well. Despite resin-based prosthetic teeth, the colour of tooth ceramic remains unchanged (Lukáts, Bujtár, Sándor, & Barabás, 2012). For osteoarticular prosthesis, zirconia enriched with yttrium oxide has been recommended to replace alumina. Greater failure strength and fatigue resistance are the key benefits. Vitreous carbon is also used because it is lighter, wear-resilient, and blood companionable. It is generally used to replace heart valves (Beckmann et al., 2015). In the form of a coating, a diamond can be used for the same purpose. Even if they are structurally and chemically equivalent to the major mineral component of bone, calcium phosphate-based ceramics are now the preferred synthetic bone material for orthopaedic and ophthalmic applications (Thamaraiselvi & Rajeswari, 2004). Porous synthetic bone substitution or scaffold materials stimulate osseointegration, including cell colonisation and revascularisation, by providing an expanded surface area (Jenkins & Grigson, 1979).

Nevertheless, contrasted to the bone, porous materials have lesser mechanical strength, making porous implants fragile. Furthermore, because the mechanical properties of ceramic materials are generally higher than that of the surrounding bone tissue, the implant may produce tensile loads at the tissue contact (Combes & Rey, 2007). Hydroxyapatite (HAP) $Ca_{10}(PO_4)6(OH)_2$, tricalcium phosphate (TCP) $Ca_3(PO_4)_2$, and combinations of HAP and TCP are common calcium phosphates found in bioceramics (Dorozhkin, 2016; Kokubo, 2008). As a result, several implant ceramics were not created with biomedical applications in mind. Nonetheless, their features and high biocompatibility have found their way into various implantable systems. For instance, TiN has been proposed as the friction surface in a hip prosthesis. However, although cell culture assays reveal satisfactory biocompatibility, implant examination reveals considerable wear due to TiN layer delamination. Another contemporary ceramic with outstanding biocompatibility that can be used in implant materials is silicon carbide (Gul, Khan, & Khan, 2020).

Biologically active ceramics have already seen particular usage responsible for biological reactivity, contrary to their typical qualities such as calcium phosphates, oxides, and hydroxides (Mala & Ruby Celsia, 2018). Bioglass and other composites combine mineral-organic lightweight structures such as HAP, alumina, or titanium dioxide with biodegradable polymers (polymethylmethacrylate): PMMA, poly(L-lactic) acid: PLLA, PL (ethylene). Bioresorbable and non-bioresorbable composites can be distinguished, resulting in the amalgamation of bioresorbable calcium phosphate (HAP) and a non-bioresorbable polymer (PMMA, PE). Due to numerous combination options and the ability to combine biological activity with mechanical qualities comparable to the bone, such materials will be even more widely used among the biomaterials (Jones & Gibson, 2020).

Bioceramics are suitable for medical use due to their anticorrosive, biocompatible, and attractive qualities. Bioinertness and non-cytotoxicity are two properties of zirconia ceramic. Carbon is another mechanical option that includes bone and blood compatibility, with neither any tissue reactivity nor cell toxicity. Bioinert ceramics do not osseointegrate, which means they do not bind with the bone. Bioactivity of bioinert ceramics, on the other hand, can be created by combining them with bioactive ceramics in composites (Choi, Conway, Cazalbou, & Ben-Nissan, 2018). Non-toxic bioactive ceramics, such as bioglasses, must develop a bone connection. The solubility of bioceramics is critical in bone repair activities like scaffolding for bone regeneration, and the slow dissolution rate of most bioceramics in contrast to bone formation rates which is still an issue in their curative utilisation (George, Reddy Peddireddy, Thakur, & Rodrigues, 2020).

Understandably, researchers are concentrating their efforts on enhancing bioceramic's dissolving characteristics while retaining or improving their mechanical capabilities. For example, glass ceramics have higher dissolving rates than crystalline materials, making them osteoinductive. On the other hand, crystalline calcium phosphate ceramics are non-toxic to tissues and have little biosorption. Because of the ceramic particle reinforcement, implant materials such as ceramic/ceramic,

ceramic/polymer, and ceramic/metal composites are now available (Somers & Lasgorceix, 2021). However, ceramic/polymer composites have discharged hazardous substances into the deeper structures. Metals are prone to corrosion, and ceramic coatings on metallic implants deteriorate when used for long periods (Dziadek, Stodolak-Zych, & Cholewa-Kowalska, 2017). Also, ceramic/ceramic composites are preferable, demonstrating biocompatibility and moulding ability. The active components in bioceramics must be tested in various in vitro and in vivo models. Performance indicators must be determined following the specific implantation site (Ranvijay Kumar, Singh, & Hashmi, 2020).

Bones serve a vital role in everyone's life, sustaining our bodies and allowing us to move in various ways. As a result, the techniques used to heal broken bones are critical, and when an area of destroyed bone is moreover extensive for self-restoration, recycled aggregates such as autografts, allografts, and synthetic materials must be used to restore the damaged bones (Abe, Kokubo, & Yamamuro, 1990). Autografts are widely used as they possess great results and are easily transferred from healthy portions of the same patient's bones. However, there appear to be problems with the limited quantity of tissue available and the additional harm to the body caused by removing the participant's bone cells (Chang et al., 2000). Bone grafting, which has been transplanted from another person, is also used, yet they have issues with limited supply, external body responses, and pathogens. As a result, artificial materials must repair bone abnormalities because they are safe and free of such limitations (Chen, Miyata, Kokubo, & Nakamura, 2000). On the other hand, synthetic materials placed in bony deformities are usually encased in fibrous tissue and do not adhere to living bone. Therefore, biologically active ceramics have garnered much interest as a solution to the foreign body reaction problem, and some of them are currently being used therapeutically as bone implants (Ohtsuki, Kamitakahara, & Miyazaki, 2009). Figure 4.1 depicts the application of ceramics as implants in the human body.

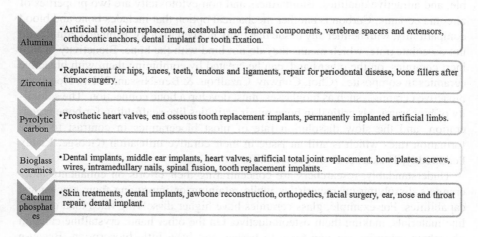

Alumina • Artificial total joint replacement, acetabular and femoral components, vertebrae spacers and extensors, orthodontic anchors, dental implant for tooth fixation.

Zirconia • Replacement for hips, knees, teeth, tendons and ligaments, repair for periodontal disease, bone fillers after tumor surgery.

Pyrolytic carbon • Prosthetic heart valves, end osseous tooth replacement implants, permanently implanted artificial limbs.

Bioglass ceramics • Dental implants, middle ear implants, heart valves, artificial total joint replacement, bone plates, screws, wires, intramedullary nails, spinal fusion, tooth replacement implants.

Calcium phosphates • Skin treatments, dental implants, jawbone reconstruction, orthopedics, facial surgery, ear, nose and throat repair, dental implant.

FIGURE 4.1 Use of ceramics as implants in the human body (Miyazaki, Kawashita, & Ohtsuki, 2016).

4.2   MATERIALS AND METHODS

The required data necessary for the analysis of content were collected from Scopus with specific keywords as shown below:

- (TITLE-ABS-KEY ("Biocompatible Ceramics") OR ("Bioactive Ceramics") AND (biomedical))

The initial keyword search resulted in 204 documents; then, the collected data were refined to English only, and 192 papers remained; then, these documents were limited to articles and resulted in 109 papers. After that, a search was limited to publication years from 2000 to 2021, and 92 articles were left. At last, the content analysis was conducted on 92 journal articles. The Biblioshiny and Vosviewer software were used to analyse the content of the articles collected on biocompatible and bioactive ceramics for biomedical applications.

4.3   CONTENT ANALYSIS OF BIOCOMPATIBLE AND BIOACTIVE CERAMICS FOR BIOMEDICAL APPLICATIONS

Content analysis is an analytical tool for determining the existence of specific words, topics, or notions in qualitative information (i.e. text). Content analysis allows researchers to quantify and analyse the incidence, connotations, and correlations of specific words, refrains, or notions. The exploration of documentation and communication artefacts, such as words in various formats, photographs, audio, and video, is known as content analysis. Social scientists use content analysis to investigate communication styles repeatedly and consistently. Contrary to replicating life interaction or gathering survey responses, one of the main advantages of employing content analysis is to analyse social phenomena that are non-invasive. Descriptive data, such as word frequencies and document lengths, can be obtained using simple computational procedures. The most basic and objective type of content analysis considers the text's unique properties, including specific words.

This chapter focuses on the content analysis of biocompatible and bioactive ceramics for biomedical applications. The main contents covered in this article are as follows:

- **Author keywords:** It covers the analysis of authors keywords employing network visualisation, word cloud of top authors keywords, and tree plot showing percentage contribution.
- **Index keyword:** It covers the analysis of index keywords employing overlay visualisation, word cloud of top index keywords, tree plot showing percentage contribution, and word dynamics from 2000 to 2021.
- **Title and abstract analysis:** It includes the title and abstract terms of articles collected from the Scopus database.

4.3.1 AUTHOR KEYWORD ANALYSIS: BIOCOMPATIBLE AND BIOACTIVE CERAMICS FOR BIOMEDICAL APPLICATIONS

Figure 4.2 shows network visualisation of author keywords in biocompatible and bioactive ceramics for biomedical applications based on the Scopus database. There are 271 author keywords, and to draw the best-suited network, the author keyword occurrence was set at two keywords. Of 271 author keywords, 46 keywords meet the desired threshold, and 41 are connected. Further, the keywords are divided into nine clusters. Figure 4.2 signifies that hydroxyapatite has a larger size compared with other keywords, and its larger bubble size indicates its maximum occurrence.

On the other hand, the plasma spraying, hap, coating, bioglass, and Fourier-transform infrared spectroscopy (FTIR) have smaller bubble sizes with different colours as these author keywords belong to different clusters, and the thickness of the lines connecting two keywords shows the strength of links between them. Hydroxyapatite seems to be the most usually applied biologically active ceramic material in surgical treatment. A bone is a synthetic tissue primarily composed of the protein keratin in its organic matrix. Collagen makes up around 10% of adult bone mass, and hydroxyapatite makes up the mineral phase. An insoluble calcium and phosphorus salt, approximately 65% of adult bone mass, is hydroxyapatite; water makes up about 25% of adult bone mass (Kumar et al., 2021; Palakurthy, Azeem, & Reddy, 2022; Singh et al., 2020).

Figures 4.3 and 4.4 represent the word cloud of the top 30 author keywords and the tree plot of author keywords. The author keyword hydroxyapatite has a significant

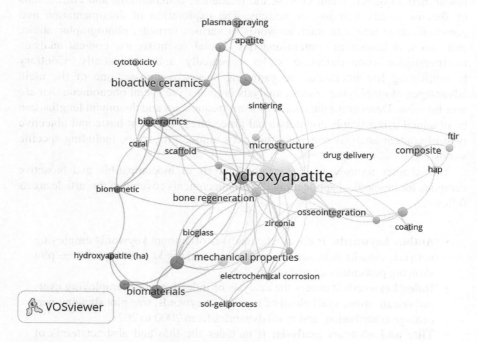

FIGURE 4.2 Author keywords: A network visualisation.

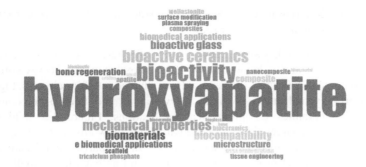

FIGURE 4.3 Top 30 author keywords: Word cloud.

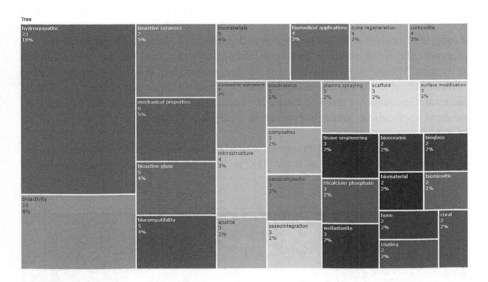

FIGURE 4.4 Author keywords: Tree plot.

role in the research as the authors provided it and appeared 23 times with a contribution of 18%. Bioactivity has appeared ten times with a contribution of 8%, followed by bioactive ceramics and mechanical properties with 5% appearance. The terms scaffolds and composite have a minor contribution. Many of the author keywords have the same meaning but are presented by the authors with little difference. Tissue engineering is a biotechnology field that uses a combination of cells, materials sciences, and appropriate biochemical and physiochemical parameters to recover, sustain, upgrade, or substitute biological tissues (Moreno Madrid, Vrech, Sanchez, & Rodriguez, 2019; Shick, Abdul Kadir, Ngadiman, & Ma'aram, 2019). The ability of materials to form a straight, persistent, and deep connection with bone tissue is referred to as bioactivity (Prakash, Singh, Pabla, & Uddin, 2018).

4.3.2 INDEX KEYWORDS ANALYSIS: BIOCOMPATIBLE AND BIOACTIVE CERAMICS FOR BIOMEDICAL APPLICATIONS

There are 1222 index keywords, and to draw the best-suited network, keyword occurrence was set at five keywords. Of 1222 index keywords, 86 keywords meet the desired threshold. Further, the keywords are divided into four clusters. Figure 4.5 shows overlay visualisation of index keywords in the field of biocompatible and bioactive ceramics for biomedical applications. Figure 4.5 signifies that hydroxyapatite has a larger size compared with other keywords, and its larger bubble size of yellow indicates its maximum occurrence. In comparison to this, animal cell, glass ceramics, sol–gel processes, biomaterials, and animals have smaller bubble sizes with different colours as these index keywords belong to different clusters, and the thickness of the lines connecting two keywords shows the strength of links between them.

Keyword indexing is an indexing methodology that classifies keywords or crucial terms in a title using natural language. Significant words are the words that have a strong relationship with the papers' actual conceptual content. Figures 4.6 and 4.7 represent the word cloud of the top 30 index keywords and the tree plot of index keywords. The index term hydroxyapatite had an essential part in the study because it was indexed in the Scopus database and occurred 59 times, accounting for 10% of the total. Ceramics occurred 38 times, accounting for 7% of all searches, followed by medical applications and biocompatibility, which accounted for 6% and 5% of all searches, respectively.

Table 4.1 shows the word dynamics related to biocompatible and bioactive ceramics in biomedical applications from 2000 to 2021 based on the data extracted from

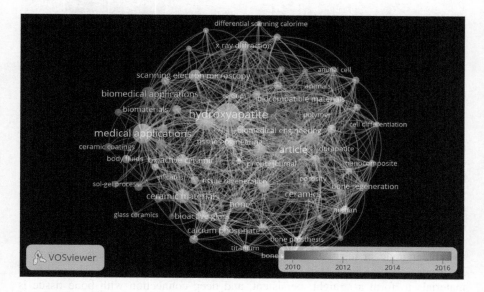

FIGURE 4.5   Index keywords: Overlay visualisation.

FIGURE 4.6   Index keyword: Word cloud of top 30.

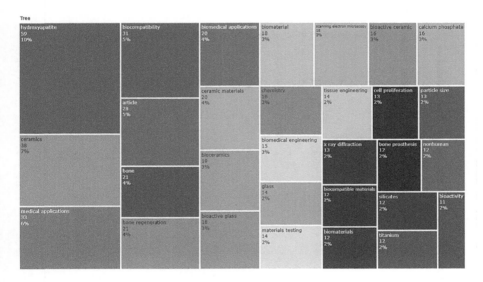

FIGURE 4.7   Tree plot: Index keywords.

Scopus, and it has a collection of articles in the journals. Hydroxyapatite has started appearing in index terms since 2001. It has appeared 582 times since its inception and occurred 59 times, maximum in 2021 with an average of 26. The word ceramic also started appearing in 2001 with a maximum of 38 appearances from 2016 to 2021. The other significant terms can be seen from Table 4.1 such as medical applications biocompatible and article followed by bone and bone regeneration. This discussion includes the top-ten word dynamics to show the most effective terms. The term bioceramics appeared at last with 181 times appearance, with an average of eight only and started appearing in 2007.

TABLE 4.1
Word Dynamics of Biocompatible and Bioactive Ceramics in Biomedical

Year	Hydroxyapatite	Ceramics	Medical Applications	Biocompatibility	Article	Bone	Bone Regeneration	Biomedical Applications	Ceramic Materials	Bioceramics
2000	0	0	1	0	0	1	0	0	1	0
2001	1	2	4	2	1	5	2	0	1	0
2002	3	4	5	2	2	5	2	0	1	0
2003	5	7	5	2	3	5	2	0	1	0
2004	6	7	5	2	3	5	2	0	2	0
2005	6	7	5	4	4	5	2	0	3	0
2006	8	8	5	4	5	5	4	0	4	0
2007	14	10	5	6	7	6	6	1	7	1
2008	15	12	5	6	9	8	8	1	7	4
2009	21	17	7	8	12	8	9	3	7	8
2010	21	17	7	8	12	8	9	3	7	8
2011	26	20	7	9	15	8	9	4	8	9
2012	28	20	8	9	16	9	9	4	8	9
2013	32	25	8	9	19	10	10	4	8	11
2014	36	28	10	12	20	10	12	5	10	12
2015	43	36	15	15	24	13	18	8	13	15
2016	50	38	18	21	25	17	21	9	16	17
2017	50	38	21	22	25	17	21	11	16	17
2018	50	38	22	23	25	18	21	12	17	17
2019	53	38	27	25	26	19	21	16	20	17
2020	55	38	30	27	27	19	21	18	20	17
2021	59	38	33	31	28	21	21	20	20	19
Total	582	448	253	247	308	222	230	119	197	181
Average	26	20	12	11	14	10	10	5	9	8
Max	59	38	33	31	28	21	21	20	20	19

4.3.3 TITLE AND ABSTRACT TERMS ANALYSIS

The "title" and "abstract" of a research paper are the first perceptions, and they must be written appropriately and precisely. Also, the title and abstract of a research paper are the most significant components and should be interesting to learn. The abstract should be consistent with the work's primary content, especially after being revised, highlighting the underlying point. It is critical to include the most important words and concepts in the title and abstract for proper indexing and retrieval from browsers and scholarly databases.

So, there is a need to explore the title and abstract terms of the specific extracted data on the biocompatible and bioactive ceramics in biomedical applications. Therefore, it is significant for new researchers to learn these applied terms. Figure 4.8 presents the density visualisation of the title and abstract terms. There are entire 2723 title and abstract terms used in the Scopus database. Of the 2723 terms, depending on a minimum of five occurrences, only 106 terms meet the threshold. Also, the most relevant terms consist of 64 terms divided into three clusters.

Table 4.2 shows the top 20 extracted title and abstract terms using VOSviewer. It shows the ranks assigned to different terms based on occurrence and relevance score. For example, the authors used the terms polymer and tissue 21 times with a relevance score of 0.5034 and 0.5173, respectively, whereas the terms implant and powder occurred 19 times with a relevance score of 0.5178 and 0.6514, respectively. On the other hand, the term Fourier ranked first on behalf of relevance score of 3.3705 but appeared only five times. Also, the successive five terms are transmission electron microscopy (TEM), electron microscopy, X-ray diffraction, FTIR, and corrosion

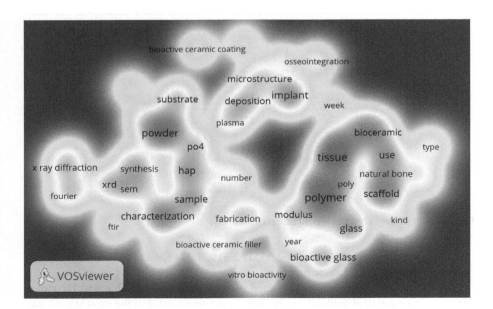

FIGURE 4.8   Title and abstract terms: Density visualisation.

TABLE 4.2

Analysis of Title and Abstract Terms

		Analysis Based on the Occurrence			Analysis Based on a Relevance Score		
Rank	Term	Occurrences	Relevance Score	Term	Relevance Score	Occurrences
1	Polymer	21	0.5034	Fourier	3.3705	5
2	Tissue	21	0.5173	TEM	2.5351	6
3	Implant	19	0.5178	Electron microscopy	2.319	8
4	Powder	19	0.6514	X-ray diffraction	2.2741	8
5	Field	17	0.4396	FTIR	1.9605	6
6	Hap	15	0.4215	Corrosion resistance	1.9497	5
7	Morphology	15	0.6279	Bone tissue engineering	1.9322	6
8	Sample	15	0.5239	Cytotoxicity	1.8757	5
9	Use	15	1.0756	Type	1.7077	7
10	Bioactive Glass	13	0.8132	XRD	1.6725	11
11	Characterisation	13	0.8876	Kind	1.6335	5
12	Glass	13	0.632	Natural Bone	1.6159	9
13	Scaffold	13	1.1469	Limitation	1.4154	5
14	Bioceramic	12	0.843	Bioactive ceramic coating	1.3931	6
15	Substrate	12	0.7748	Clinical application	1.2836	10
16	Approach	11	0.7169	Vivo	1.172	5
17	Concentration	11	0.6758	Scaffold	1.1469	13
18	Deposition	11	0.3614	SEM	1.1019	9
19	Modulus	11	0.4099	Use	1.0756	15
20	Paper	11	1.0281	Vitro bioactivity	1.0659	6

FIGURE 4.9   Word cloud of top 30 title terms.

resistance with a relevance score of 2.5351, 2.319, 2.2741, 1.9605, and 1.9497, having an occurrence of 6, 8, 8, 6, and 5.

The top 30 title terms of the extracted data on the biocompatible and bioactive ceramics in biomedical applications are shown in Figure 4.9, and the most significant top 30 abstract terms are shown in Figure 4.10.

FIGURE 4.10   Word cloud of top 30 abstract terms.

4.4   CONCLUDING REMARKS ON CONTENT ANALYSIS

According to the content analysis of author keywords connected to biocompatible and bioactive ceramics for biomedical applications, hydroxyapatite has the highest incidence. Plasma spraying, hap, coating, bioglass, and FTIR are some of the additional terms with lesser size. The author keyword hydroxyapatite had a big part in the study because it was contributed by the authors and occurred 23 times, accounting for 18% of the total. Bioactivity has featured ten times, accounting for 8% of the total, followed by bioactive ceramics and mechanical characteristics, each accounting for 5%. Hydroxyapatite has an immense contribution to index keyword analysis. Animal cells, glass ceramics, sol–gel processes, biomaterials, and mammals, on the other hand, have smaller bubble diameters. Because it was indexed in the Scopus database and appeared 59 times, accounting for 10% of the total, the index word hydroxyapatite played an essential role in the study. Ceramics came up 38 times, accounting for 7% of all searches, followed by medical applications and biocompatibility, accounting for 6% and 5% of all searches, respectively. Since 2001, the word hydroxyapatite has begun to emerge in index terms. Since its beginning, it has appeared 582 times, with a maximum of 59 times in 2021 and an average of 26. The term bioceramics first appeared in 2007 and has since appeared 181 times with an average of eight times. There are 2723 title and abstract phrases based on a minimum of five occurrences. The terms polymer and tissue were mentioned 21 times, with a relevance score of 0.5034 and 0.5173, respectively, whereas implant and powder were mentioned 19 times, with a relevance value of 0.5178 and 0.6514, respectively. The term Fourier, on the other hand, came in top place with a relevance score of 3.3705 but only featured five times. TEM, electron microscopy, X-ray diffraction, FTIR, and corrosion resistance are the successive five terms, with a relevance score of 2.5351, 2.319, 2.2741, 1.9605, and 1.9497, respectively, and an occurrence of 6, 8, 8, 6, and 5.

4.5 FUTURE DIRECTIONS AND OUTLOOKS

Any element, natural or artificial, that treats augments or replaces any tissue, organ, or biological function is referred to as a biomaterial. Because of the critical criteria and biocompatibility, biomaterial selection is one of the most challenging concerns, and it has piqued the interest of material scientists in recent years. Metals are vulnerable to corrosion, resulting in the release of byproducts that might induce undesirable biological reactions. Because of their biocompatibility, ceramics are appealing as biological implants. The high-mechanical-strength alumina has minimal tissue reactivity, is harmless to tissues, and passes blood compatibility testing. Carbon with mechanical qualities comparable to bone is a promising contender because it is blood compatible, has no tissue reaction, and is harmless to cells. Bioengineering and controlled drug release technologies have significantly benefited from the availability of a diverse spectrum of polymers. Appreciating advancements in biocomposites design and production processes, implants with higher performance are becoming more possible (Brunello et al., 2020; Durairaj et al., 2020; Li et al., 2020).

Nevertheless, surgeons must be convinced of composite biomaterials' protracted performance and reliability for successful application. The success of materials in biomedical applications was the product of serendipity, ongoing improvement in manufacturing technique, and breakthroughs in material surface modification, rather than painstaking decisions based on biocompatibility considerations. Numerous parameters in the coming generations must support the selection of a biomaterial for a specific application. Each biomaterial must meet an essential requirement of biocompatibility. At the molecular/genetic level, medical research wants to investigate new scientific frontiers for detecting, healing, alleviating, and preventing illness (Liu, Gong, Liu, Long, & Dong, 2020; Pawlik, Dziadek, Cholewa-Kowalska, & Osyczka, 2020).

The selection of acceptable materials with excellent mechanical properties is the first and most important prerequisite. Biomaterials should be biocompatible, bioresorbable, biodegradable at a controlled rate, have high mechanical strength, and be bioactive. The next task is to create scaffolds with appropriate structures. To develop medical implants and technologies, bone-tissue engineering also used a variety of forms and architectures. Their design heavily influences the mechanical properties of scaffold structures (Alizadeh-Osgouei, Li, & Wen, 2019). Three-dimensional porous scaffolds enable rapid cell growth and adhesion and high nutrition and waste transfer rates. Such structures also provide a lot of surface area for bone development (Kamboj, Ressler, & Hussainova, 2021; Klyui et al., 2021). As a result, highly porous scaffolds in 3D forms are critical in achieving these goals. Scaffolds are recommended to have a porosity of about 90%. Another important criterion that should be investigated further is the adherence of coatings to substrates. Efforts have been undertaken to improve adhesive strength in a variety of ways. By forming solid chemical connections between HAP nanoparticles and the PLLA and PLGA matrix, hydrogel groups of hydroxyapatite (HA), such instance, can boost adhesiveness. More importantly, HA-containing composites have higher mechanical strength than pure composites (Guarino et al., 2020).

A further critical need for biodegradable materials is that they have appropriate degradation behaviour. The addition of nanofillers to a biodegradable composite could slow down the pace of deterioration. Polymeric materials are being used as a modern trend of restorative material due to their exceptional and favourable properties (Ohtsuki et al., 2009). Even though several biomaterials have been thoroughly explored for biological applications, many more materials might be investigated. In this regard, the physicochemical characteristics of this new generation of biomaterials and their morphologies and potential applications in tissue engineering should be explored further. It is hoped that the identified issues will be mitigated by using polymers. Traditional production methods may not yield biomaterials that mimic the structure of bone (Arun Prakash, Venda, Thamizharasi, & Sathya, 2021; Ferreira et al., 2021). On the other hand, innovative approaches restrict raw resources and outputs. For example, only a limited variety of materials, such as thermosets, are currently available for use in 3D-printing equipment, and these materials do not meet all of the standards (Alizadeh-Osgouei et al., 2019).

In addition, achieving strong interfacial adhesion is difficult due to the varied forms of scaffolds. A range of engineering and production procedures have been researched in medical science to develop long-lasting implant devices with good biological properties, good biocompatibility, and bioactivity. However, developing a variety of better biomaterials and production techniques remains critical (Safavi, Surmeneva, Surmenev, & Khalil-Allafi, 2021; Tang et al., 2021).

REFERENCES

Abe, Y., Kokubo, T., & Yamamuro, T. (1990). Apatite coating on ceramics, metals and polymers utilizing a biological process. *Journal of Materials Science: Materials in Medicine, 1*(4), 233–238. doi:10.1007/BF00701082

Alizadeh-Osgouei, M., Li, Y., & Wen, C. (2019). A comprehensive review of biodegradable synthetic polymer-ceramic composites and their manufacture for biomedical applications. *Bioactive Materials, 4*, 22–36. doi:10.1016/j.bioactmat.2018.11.003

Arun Prakash, V. C., Venda, I., Thamizharasi, V., & Sathya, E. (2021). A new attempt on synthesis of spherical nano hydroxyapatite powders prepared by dimethyl sulfoxide - poly vinyl alcohol assisted microemulsion method. *Materials Chemistry and Physics, 259*. doi:10.1016/j.matchemphys.2020.124097

Beckmann, N. A., Gotterbarm, T., Innmann, M. M., Merle, C., Bruckner, T., Kretzer, J. P., & Streit, M. R. (2015). Long-term durability of alumina ceramic heads in THA. *BMC Musculoskeletal Disorders, 16*(1), 249. doi:10.1186/s12891-015-0703-2

Brunello, G., Biasetto, L., Elsayed, H., Sbettega, E., Gardin, C., Scanu, A.,. Sivolella, S. (2020). An in vivo study in rat femurs of bioactive silicate coatings on titanium dental implants. *Journal of Clinical Medicine, 9*(5). doi:10.3390/jcm9051290

Chang, B.-S., Hong, K.-S., Youn, H.-J., Ryu, H.-S., Chung, S.-S., & Park, K.-W. (2000). Osteoconduction at porous hydroxyapatite with various pore configurations. *Biomaterials, 21*(12), 1291–1298. doi:10.1016/S0142-9612(00)00030-2

Chen, Q., Miyata, N., Kokubo, T., & Nakamura, T. (2000). Bioactivity and mechanical properties of PDMS-modified CaO–SiO2–TiO2 hybrids prepared by sol-gel process. *Journal of Biomedical Materials Research, 51*(4), 605–611. doi:10.1002/1097-4636 (20000915)51:4<605::AID-JBM8>3.0.CO;2-U

Choi, A. H., Conway, R. C., Cazalbou, S., & Ben-Nissan, B. (2018). 3 - Maxillofacial bio-ceramics in tissue engineering: Production techniques, properties, and applications. In S. Thomas, P. Balakrishnan, & M. S. Sreekala (Eds.), *Fundamental biomaterials: Ceramics* (pp. 63–93). Woodhead Publishing. ISBN 9780081022030; https://doi.org/10.1016/B978-0-08-102203-0.00003-2. https://www.sciencedirect.com/science/article/pii/B9780081022030000032

Combes, C., & Rey, C. (2007). Bioceramics. In *Ceramic materials* (pp. 493–521).

Doremus, R. (1992). Bioceramics. *Journal of Materials Science, 27*(2), 285–297.

Dorozhkin, S. V. (2016). Multiphasic calcium orthophosphate (CaPO4) bioceramics and their biomedical applications. *Ceramics International, 42*(6), 6529–6554. doi:10.1016/j.ceramint.2016.01.062

Durairaj, R. B., Nagaraj, M., Mageshwaran, G., Joesph, G. B., Sriram, V., & Jeevahan, J. (2020). Effect of plasma sprayed α-tri-calcium phosphate (α- TCP)deposition over metal-lic biomaterial surfaces for biomedical applications. *Digest Journal of Nanomaterials and Biostructures, 15*(2), 345–358.

Dziadek, M., Stodolak-Zych, E., & Cholewa-Kowalska, K. (2017). Biodegradable ceramic-polymer composites for biomedical applications: A review. *Materials Science and Engineering: C, 71*, 1175–1191. doi:10.1016/j.msec.2016.10.014

Eliaz, N., & Metoki, N. (2017). Calcium phosphate bioceramics: A review of their history, structure, properties, coating technologies and biomedical applications. *Materials, 10*(4), 334.

Fabbri, M., Celotti, G. C., & Ravaglioli, A. (1995). Hydroxyapatite-based porous aggregates: Physico-chemical nature, structure, texture and architecture. *Biomaterials, 16*(3), 225–228. doi:10.1016/0142-9612(95)92121-L

Ferreira, F. V., Otoni, C. G., Lopes, J. H., de Souza, L. P., Mei, L. H. I., Lona, L. M. F.,. Mattoso, L. H. C. (2021). Ultrathin polymer fibers hybridized with bioactive ceramics: A review on fundamental pathways of electrospinning towards bone regeneration. *Materials Science and Engineering C, 123.* doi:10.1016/j.msec.2020.111853

George, A. M., Reddy Peddireddy, S. P., Thakur, G., & Rodrigues, F. C. (2020). Chapter 29 - Biopolymer-based scaffolds: Development and biomedical applications. In K. Pal, I. Banerjee, P. Sarkar, D. Kim, W.-P. Deng, N. K. Dubey, & K. Majumder (Eds.), *Biopolymer-based formulations* (pp. 717–749). Elsevier.

Guarino, V., Iafisco, M., & Spriano, S. (2020). 1 - Introducing biomaterials for tissue repair and regeneration. In V. Guarino, M. Iafisco, & S. Spriano (Eds.), *Nanostructured biomaterials for regenerative medicine* (pp. 1–27). Woodhead Publishing.

Gul, H., Khan, M., & Khan, A. S. (2020). 3 - Bioceramics: Types and clinical applications. In A. S. Khan, & A. A. Chaudhry (Eds.), *Handbook of ionic substituted hydroxyapatites* (pp. 53–83): Woodhead Publishing.

Hench, L. L. (1993). *An introduction to bioceramics* (Vol. 1). World Scientific.

Jenkins, G., & Grigson, C. (1979). The fabrication of artifacts out of glassy carbon and carbon-fiber-reinforced carbon for biomedical applications. *Journal of Biomedical Materials Research, 13*(3), 371–394.

John, A. A., Subramanian, A. P., Vellayappan, M. V., Balaji, A., Jaganathan, S. K., Mohandas, H., ... Yusof, M. (2015). Review: Physico-chemical modification as a versatile strategy for the biocompatibility enhancement of biomaterials. *RSC Advances, 5*(49), 39232–39244. doi:10.1039/C5RA03018H

Jones, J. R., & Gibson, I. R. (2020). 1.3.4 - ceramics, glasses, and glass-ceramics: Basic principles. In W. R. Wagner, S. E. Sakiyama-Elbert, G. Zhang, & M. J. Yaszemski (Eds.), *Biomaterials science* (Fourth Edition) (pp. 289–305). Academic Press.

Kamboj, N., Ressler, A., & Hussainova, I. (2021). Bioactive ceramic scaffolds for bone tissue engineering by powder bed selective laser processing: A review. *Materials, 14*(18). doi:10.3390/ma14185338

Kinnari, T. J., Esteban, J., Gomez-Barrena, E., Zamora, N., Fernandez-Roblas, R., Nieto, A., Vallet-Regí, M. (2009). Bacterial adherence to SiO2-based multifunctional bioceramics. *Journal of Biomedical Materials Research Part A: An Official Journal of The Society for Biomaterials, The Japanese Society for Biomaterials, and The Australian Society for Biomaterials and the Korean Society for Biomaterials*, *89*(1), 215–223. doi:10.1002/jbm.a.31943

Klyui, N. I., Chornyi, V. S., Zatovsky, I. V., Tsabiy, L. I., Buryanov, A. A., Protsenko, V. V., Gryshkov, O. (2021). Properties of gas detonation ceramic coatings and their effect on the osseointegration of titanium implants for bone defect replacement. *Ceramics International*, *47*(18), 25425–25439. doi:10.1016/j.ceramint.2021.05.265

Kokubo, T. (2008). *Bioceramics and their clinical applications*. Elsevier.

Kumar, R., Dubey, R., Singh, S., Singh, S., Prakash, C., Nirsanametla, Y., ... Chudy, R. (2021). Multiple-criteria decision-making and sensitivity analysis for selection of materials for knee implant femoral component. *Materials*, *14*(8). doi:10.3390/ma14082084

Kumar, R., Singh, R., & Hashmi, M. S. J. (2020). Polymer- ceramic composites: A state of art review and future applications. *Advances in Materials and Processing Technologies*, 1–14. doi:10.1080/2374068X.2020.1835013

Li, R. T., Li, Z., Hu, H. L., Liu, Z. Q., Wang, Y., & Khor, K. A. (2020). Nacre-like Co–Cr–Mo/Ti2O3 coating on the Co–Cr–Mo substrate prepared using spark plasma sintering. *Ceramics International*, *46*(8), 10530–10535. doi:10.1016/j.ceramint.2020.01.054

Liu, S. Y., Gong, W. Y., Liu, M. Q., Long, Y. Z., & Dong, Y. M. (2020). Clinical efficacy observation of direct pulp capping using iRoot BP Plus therapy in mature permanent teeth with carious pulp exposure. *Zhonghua kou qiang yi xue za zhi = Zhonghua kouqiang yixue zazhi = Chinese Journal of Stomatology*, *55*(12), 945–951. doi:10.3760/cma.j.cn112144-20200327-00173

Lukáts, O., Bujtár, P., Sándor, G. K., & Barabás, J. (2012). Porous hydroxyapatite and aluminium-oxide ceramic orbital implant evaluation using CBCT scanning: A method for <i>in vivo</i> porous structure evaluation and monitoring. *International Journal of Biomaterials*, *2012*, 764749. doi:10.1155/2012/764749

Mala, R., & Ruby Celsia, A. S. (2018). 8 - Bioceramics in orthopaedics: A review. In S. Thomas, P. Balakrishnan, & M. S. Sreekala (Eds.), *Fundamental biomaterials: Ceramics* (pp. 195–221). Woodhead Publishing.

Miyazaki, T., Kawashita, M., & Ohtsuki, C. (2016). Ceramic-polymer composites for biomedical applications. In *Handbook of bioceramics and biocomposites* (pp. 287–300). Springer International Publishing.

Moreno Madrid, A. P., Vrech, S. M., Sanchez, M. A., & Rodriguez, A. P. (2019). Advances in additive manufacturing for bone tissue engineering scaffolds. *Materials Science and Engineering: C*, *100*, 631–644. doi:10.1016/j.msec.2019.03.037

Ohtsuki, C., Kamitakahara, M., & Miyazaki, T. (2009). Bioactive ceramic-based materials with designed reactivity for bone tissue regeneration. *Journal of the Royal Society, Interface*, *6 Suppl 3*(Suppl 3), S349–S360. doi:10.1098/rsif.2008.0419.focus

Palakurthy, S., Azeem, P. A., & Reddy, K. V. (2022). Sol–gel synthesis of soda lime silica-based bioceramics using biomass as renewable sources. *Journal of the Korean Ceramic Society*, *59*(1), 76–85. doi:10.1007/s43207-021-00163-z

Park, J. (2009). *Bioceramics: properties, characterizations, and applications* (Vol. 741). Springer Science & Business Media.

Pawlik, J., Dziadek, M., Cholewa-Kowalska, K., & Osyczka, A. M. (2020). The synergistic effect of combining the bioactive glasses with polymer blends on biological and material properties. *Journal of the American Ceramic Society*, *103*(8), 4558–4572. doi:10.1111/jace.17131

Prakash, C., Singh, S., Pabla, B. S., & Uddin, M. S. (2018). Synthesis, characterization, corrosion and bioactivity investigation of nano-HA coating deposited on biodegradable Mg-Zn-Mn alloy. *Surface and Coatings Technology*, *346*, 9–18. doi:10.1016/j.surfcoat.2018.04.035

Safavi, M. S., Surmeneva, M. A., Surmenev, R. A., & Khalil-Allafi, J. (2021). RF-magnetron sputter deposited hydroxyapatite-based composite & multilayer coatings: A systematic review from mechanical, corrosion, and biological points of view. *Ceramics International*, *47*(3), 3031–3053. doi:10.1016/j.ceramint.2020.09.274

Semlitsch, M., Lehmann, M., Weber, H., Doerre, E., & Willert, H. G. (1977). New prospects for a prolonged functional life-span of artificial hip joints by using the material combination polyethylene/aluminium oxide ceramic/metal. *Journal of Biomedical Materials Research*, *11*(4), 537–552. doi:10.1002/jbm.820110409

Shick, T. M., Abdul Kadir, A. Z., Ngadiman, N. H. A., & Ma'aram, A. (2019). A review of biomaterials scaffold fabrication in additive manufacturing for tissue engineering. *Journal of Bioactive and Compatible Polymers*, *34*(6), 415–435. doi:10.1177/0883911519877426

Singh, G., Singh, S., Prakash, C., Kumar, R., Kumar, R., & Ramakrishna, S. J. P. C. (2020). Characterization of three-dimensional printed thermal-stimulus polylactic acid-hydroxyapatite-based shape memory scaffolds. *Polymer Composites*, *41*(9), 3871–3891. doi:10.1002/pc.25683

Somers, N., & Lasgorceix, M. (2021). Surface treatment of bioceramics. In M. Pomeroy (Ed.), *Encyclopedia of materials: Technical ceramics and glasses* (pp. 701–715). Elsevier.

Tang, Q., Li, X., Lai, C., Li, L., Wu, H., Wang, Y., & Shi, X. (2021). Fabrication of a hydroxyapatite-PDMS microfluidic chip for bone-related cell culture and drug screening. *Bioactive Materials*, *6*(1), 169–178. doi:10.1016/j.bioactmat.2020.07.016

Thamaraiselvi, T., & Rajeswari, S. (2004). Biological evaluation of bioceramic materials-a review. *Carbon*, *24*(31), 172.

Williams, D. F. (2008). On the mechanisms of biocompatibility. *Biomaterials*, *29*(20), 2941–2953. doi:10.1016/j.biomaterials.2008.04.023

5 Stimuli Responsive Bio-Based Hydrogels
Potential Employers for Biomedical Applications

Lalita Chopra and Manikanika
Chandigarh University, Mohali, India

Jasgurpreet Singh Chohan
University Centre for Research and Development,
Chandigarh University, Mohali, India

CONTENTS

DOI: 10.1201/9781003266464-5

5.1 INTRODUCTION

Bio-based polymers mean polymeric potential materials that are produced or extracted from the available or many times waste biomass in the nature. Our mother nature is full of biomass and many times its potential is underestimated (Nakajima et al., 2017). If I talk about polymer, it's now a days a very old term but not fully exhausted yet. Polymeric materials can find uses all over the world and the best way to find their application is to see our surroundings such as our clothing, household materials, support material, automobile, and paper. Maximum of these polymers are synthetic ones; these are no doubt useful but yet have many disadvantages such as non-biodegradability, toxicity, long exposure side effect, and environmental concern (Garrison et al., 2016). Whereas biopolymers here come up with all benefits such as biodegradability, green, zero waste, non-toxicity, and environmental friendly. Therefore, now a days, researchers have widely shifted towards the exploration of biopolymers because of their superiority among the synthetic polymers. A huge number of biopolymers such as dextrin, agar-agar, cellulose, hemicellulose, chitin, and chitosan were known in this regard (Muñoz-Bonilla et al., 2019). These biopolymers are biocompatible to the living tissues and interact with the biomolecules in such a way that these will find utilisation in the carriage and delivery of biomolecules, tissue repair, tissue culture, wound dressing, ophthalmology, and biowaste removal (Kawaguchi et al., 2017). So, their applications are numerous, and all credit goes to the unique properties of the bio-based polymeric systems. For in vivo applications, more emphasis is given to the materials that are responsive to certain stimuli or the presence or absence of the stimulus decides the activity of the material (Nath & Chilkoti, 2002). Stimulus can be a chemical or environmental such as time, pH, temperature, agent, and solvent. Generally, biopolymers are porous, hydrogel like materials; therefore, they show swelling–deswelling behaviour in the presence and absence of stimuli (De las Heras Alarcón et al., 2005). Upon swelling they can imbibe the material in themselves and when deswelling occurs the material will be released in the required place. Therefore, they work like biotransport in the body (Sheth et al., 2013). Dina M. Silva and his co-workers worked with bio-based hydrogels consisting of dextrin nanogels along with the urinary bladder matrix so that the biocompatibility of the synthesized hydrogels was attested by cell encapsulation. The synthesised hydrogels showed properties suitable for cell therapy and regenerative medicine (Silva et al., 2014). K. Senthilarasan et al. worked on tissue engineering and bone regeneration hydrogels synthesized by using nano hydroxyapatite with agar–agar (Velusamy et al., 2021). Haishun Du and co-workers published a review on the progress in the field of preparation of hydrogels based on cellulose nanocrystals (CNCs) and cellulose nanofibrils (CNFs) and their horizon of applications in biomedical activities such as tissue scaffold engineering, wound dressing, and drug delivery applications (Du et al., 2019).

Biopolymers have many limitations such as solubility, less porosity, and less efficiency. These limitations can be removed by modifying some physical or chemical technique (Sheth et al., 2013). Polymeric blends are also gaining more and more attention because of enhancement of the properties of the biopolymers by fabricating these along with some monomers so as to inculcate the required functionalities in them (Wang et al., 2020; Fertahi et al., 2021). Muhammad Aamir Sajid modified structures of chitosan and chitin biopolymer upon subjecting it to processes such as N-phthaloylation, acylation, alkylation, phosphorylation O-carboxymethylation, Schiff base formation, N-carboxyalkylation, quaternisation, graft copolymerisation, and sulfonation to resolve the limitations of the biopolymers and to increase their efficiency in biomedical applications (Sajid et al., 2018). Similarly, Ganeswar Dalei fabricated smart hydrogels by encompassing carboxymethyl guar gum and chitosan (CMGG/CS). The improvement in the biological efficacy was gained by non-thermal plasma-assisted modifications by the use of O_2, air, argon gas, etc. (Dalei et al., 2019).

Here, this review represents the interaction of bio-based hydrogels with the biomolecules and hence the applicability of these biopolymers in biomedical field and future prospective as well.

5.2 HYDROGELS

Bio-based polymeric materials are ultimate candidates with hydrogel-like properties. A hydrogel is a material which when immersed in a water-based system imbibes water 500 times more of its weight and also is capable of holding it for a longer time. That is why hydrogel materials were used for the internal lining of baby's nappies, hygienic pads, etc. (Ajdary et al., 2021). Along with water, these can imbibe even the biological constituents such as drugs, proteins, RNA/DNA, hormones, and enzymes and can act as delivery agents for in vivo and in vitro studies as well (Mondal et al., 2020). The work on the hydrogel field had started in 1960 and the main credit goes to Wichterle and Lim who shifted the focus of the researchers towards multifunctional and multitalented 3D networks (© *1960 Nature Publishing Group*, 1960). Hydrogels exhibited excellent biocompatibility with the biological tissues as they can be thought to be the replacement of these in many fields of bio-engineering. This biocompatibility explored their applicability and expectations in the biomedical field. Hydrogels are composed of a number of hydrophilic functional groups such as –OH, –COOH, –NH2, and –CONH2 and are also considered as colloidal gels, with water being the dispersion solution (Huang et al., 2020) (Figure 5.1).

These are exceedingly absorbents and the hydrogels utilised for wound dressing, bandages, dissolve and are finally submerged with the human skin (Chai et al., 2017). So, they carry healing properties too. Numerous researchers called them intelligent or smart materials being sensitive to the changes in their surroundings and act accordingly. The gels have properties such as swelling, porosity, crosslinking, and mechanical strength (Peppas et al., 2006).

5.3 SWELLING

One of the foremost characteristics of the hydrogels is to absorb water (when immersed in it) and also to retain it in itself for a longer time. Second one is to release water slowly in a constant way. Hydrogels can absorb from 10% (roughly lowest one) to 1000 times than their own body weight and hence swelled up gently. This property of swelling is owned to the network-like structure and the presence of

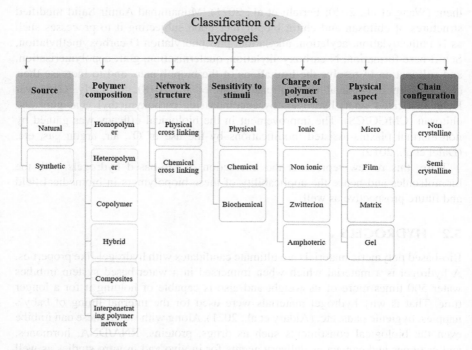

FIGURE 5.1   Classification of hydrogels.

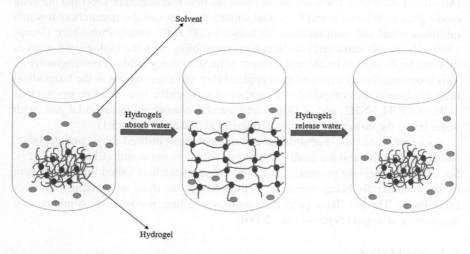

FIGURE 5.2   Swelling–deswelling phenomenon of hydrogels.

large number of hydrophilic functional groups on to the surface (Can et al., 2007). Swelling depends upon the hydrogel–solvent interaction and evaluation of swelling is the major criterion to judge the performance of a hydrogel (Figure 5.2).

The work on the swelling behaviour of the hydrogels explored a lot in the 1990s with the discovery of hydrogels. Swelling also decides the potency of the hydrogel

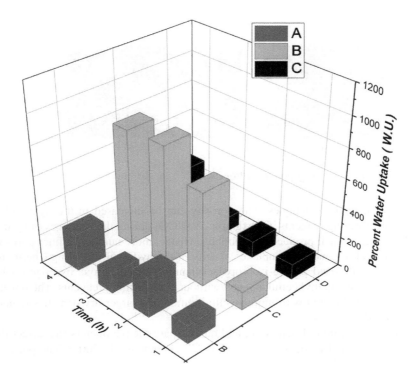

FIGURE 5.3   Representing % W.U. with respect to contact time in hours.

with respect to the stimulus. The swelling depends upon the weight of the swollen polymer and the weight of the dry polymer (Chen et al., 1999). The per cent water uptake of per cent swelling can be presented as:

$$\% \text{ W.U. or } \% \text{ Swelling} = \frac{\text{Weight of the swollen polymer} - \text{Weight of dry polymer}}{\text{Weight of dry polymer}} \times 100$$

Swelling investigations can be performed with respect to time (e.g., 1 hour to 24 hours) or other factors such as pH, temperature, and ionic strength (Dolbow et al., n.d.). A graph can be plotted between % W.U. verses the factor against which swelling is represented, which is shown in Figure 5.3.

5.4   POROSITY

Porosity is a morphological characteristic of the material which determines the presence of void surface. Hydrogels are porous structures. The surface morphology of the hydrogels and the presence, abundance, and homogeneity of the pores on the surface can be confirmed by the scanning electron microscope and transmission electron

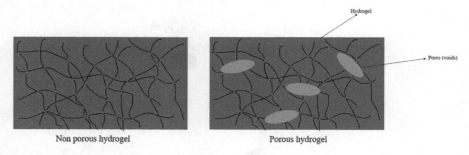

FIGURE 5.4   Porous vs non-porous hydrogels.

microscope analysis techniques. The porosity can be increased in the hydrogel if we add pore-forming reagents, for example, $NaHCO_3$ during the synthesis of hydrogels. These agents will produce CO_2 bubbles. Hence, more number of finer pores in turn affect swelling and water uptake capacity of the hydrogel. Jia-Ming Chern and co-workers synthesised poly(n-isopropylacrylamide)-based porous hydrogels with ($NaHCO_3$, carboxymethyl cellulose) and without pore-forming reagents. The porous hydrogels were obtained with more swelling ration and larger dry densities (Chern et al., 2004) (Figure 5.4).

The pore sizes, pore distribution, and quantity are the parameters that affect the performance of the hydrogels directly (Van Vlierberghe et al., 2007). The porosity percentage can be determined as:

$$\% \text{ Porosity} = \frac{\text{Volume of pores}}{\text{Volume of bulk} + \text{Volume of pores}} \times 100$$

5.5   CROSSLINKING

Crosslinking is one of the characteristics of the hydrogels which in turn is related with other characteristics such as porosity, swelling, and mechanical strength. The degree of crosslinking can vary from one to another hydrogel. The hydrogels can be crosslinked by physical techniques as well by the chemical methods (Hennink & van Nostrum, 2012). Crosslinking can convert a simple chain structure into a three-dimensional interwoven structure. This structure will help in efficient immobilisation as well as release of biologically active species. The crosslinking can be achieved by a large number of techniques such as chemical crosslinking by using crosslinkers such as glutaraldehyde, N,N′-methylene bis acrylamide (N,N′-MBA), glyoxal, ethylene glycol di methacrylate (EGDMA), epichlorohydrin, etc. (Maitra & Shukla, 2014). The crosslinking can also be achieved by physical processes such as radiation-mediated crosslinking, microwave-assisted crosslinking and by mechanical methods as well (Figure 5.5).

Natural polymers were crosslinked for the enhancement in the network-like structure to be applied in the fields of drug delivery, paper industry, agricultural practices, indoor plantation, scaffolds, etc. Table 5.1 represents the modification of

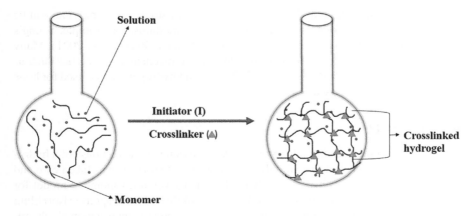

FIGURE 5.5   Crosslinking of linear chains by using crosslinker.

TABLE 5.1
Crosslinking of Biopolymers for Highly Efficient Biomedical Applications

Hydrogel	Crosslinker	Efficacy in Biomedical Field	Reference
Starch	Glutaraldehyde	Wound healing and tissue engineering	Manuscript (2019)
Dextrin	N,N'-methylene bis acrylamide (N,N'-MBA)	Drug delivery applications	Das et al. (2014)
Chitosan	Genipin	Three-dimensional scaffolds for liver tissue engineering applications	Zhang et al. (2016)
Chitin	Epichlorohydrin	Good biocompatibility and breathable properties	Zhu et al. (2019)
Cellulose	1,2,3,4-Butanetetracarboxylic acid dianhydride	Hydrogels exhibited good biodegradability, with a maximum degradation of 95% within seven days using cellulose	Kono & Fujita (2012)
Gelatine	Glyoxal	Bone regeneration	Wang & Stegemann (2011)
Guar gum	Epichlorohydrin	Property enhancement for biomedical applications	Hongbo et al. (2012)

bio-based hydrogels using different crosslinking agents so as to enhance their efficacy in the biomedical field and other fields.

5.6   MECHANICAL STRENGTH

Mechanical strength of hydrogels can be their strength to withstand stress and strain. Mechanical properties can vary with respect to the degree of crosslinking. The work on the mechanical properties of the hydrogels has started in the early 1990s. Hydrogels can be fabricated in such a way so as to get desirable mechanical properties in terms of stiffness and crosslinking. The mechanical strength of a polymer can be measured in terms of Young's modulus or Poisson's modulus, for example, Young's modulus of silk fibres is too high than that of agar–agar (Zhao et al., 2021). Many mechanisms were employed in order to enhance the mechanical properties such as fibrous reinforcement by Beckett et al. (2020). Strong hydrogels can be used for bone replacements and tissue scaffolds.

5.7 STIMULI

Since the past three decades, a lot of work has been done on the response of a hydrogel or biological material to a stimulus. The concept of stimulus was taken care and well explained in the 1990s. Sylvan Kornblum had given a taxonomical model for the cognitive basis of the stimuli as well as compatibility of the response (Kornblum et al., 1990). What is stimulus/stimuli? Stimulus can be any environmental, chemical, biochemical, mechanical, optical signal, or agent which triggers up the behaviour of a hydrogel in a specific way. Stimulus can be physical or chemical. Physical stimulus includes temperature, pressure, magnetic, and electrical field. Chemical stimulus includes pH, enzymes, solvents alteration, ionic strength, and biomolecule (Wilson & Guiseppi-Elie, 2013). Similarly, hydrogels can be exposed to a stimulus for understanding their working in vivo and in vitro conditions. Before performing in vivo experimentations, it is suggested to check the behavioural changes in the hydrogel in the presence of stimuli under in vitro conditions first and if after batched of experiments suits well then go for in vivo conditions. Many researchers are now exploring the use of the hydrogels for responding to multiple stimuli, for example, Ali Pourjavadi et al. prepared multiresponsive hydrogels to be used in in vivo or in vitro milieu (Pourjavadi et al., 2021) (Figure 5.6).

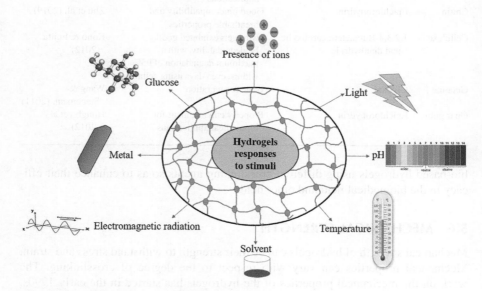

FIGURE 5.6   Stimuli to which hydrogel respond.

FIGURE 5.7 Representation of porous, homogenous structure of hydrogel.

5.8 TYPES OF HYDROGELS ON THE BASIS OF STIMULI

Hydrogel is one of the wide terms that came into existence past half a century. The interest started with the fabrication of the hydrogels of HEMA (hydroxyethyl methacrylate) monomer by crosslinking it. But now a days, hydrogel has received a fascinating position among the scientists such as bio-scientists and industrial scientists and also among environmentalists (Chai et al., 2017) (Figure 5.7).

Hydrogels are the material class which is sensitive, smart, intelligent, self-sustaining, and self-supporting. These are the gels that consist of a 3D network-like structure which allows in and out diffusion of various bio and organic molecules. In 1984, Lee and co-workers described that hydrogel is any substance which resulted in the formation of colloidal gel. Hydrogels can be classified into a number of ways such as the source of hydrogel, composition of polymer, network structure, charge on hydrogel, physical aspects, and stimuli responsiveness (Li et al., 2004). Stimuli-responsive hydrogels have found more interest due to their behaviours of showing remarkable change in their properties in contact to the particular stimulus. Due to these behaviours, they are known to be smart hydrogels and find applications in almost every field especially the biomedical field (Pimenta et al., 2019). Many researchers have done remarkable work in terms of responsiveness of the hydrogels for a specific stimulus.

5.9 PH RESPONSIVE

pH-responsive hydrogels are those that are sensitive to the change in pH. In the biomedical field, pH plays an important role as there is quite a variation in pH in the human body and other body organs and fluids. The gastrointestinal tract of human beings has a variation in pH from most acidic to most basic. pH-responsive hydrogels show a swelling–deswelling behaviour with variation in pH. These also show the highest swelling at a particular pH (Gupta et al., 2002). Therefore, they can imbibe solvent up to the maximum and after deswelling they can release the solvent.

James P. Best and his fellow co-workers fabricated thiol-modified poly(methacrylic acid) hydrogel capsules and experimented swelling at pH 7.4 to 4.0. The capsule becomes stiffer at pH 4.0 (Best et al., 2013). Sudipto K. De understood the swelling–deswelling kinetic behaviour of the hydrogels by applying kinetic models such as chemical diffusion equation and mechanical equations (De et al., 2002). Some hydrogels shows swelling at acidic pH, some at alkaline pH, for example, chitosan consisting of $-NH_2$, $-OH$ functionalities shows more swelling in acidic medium, carboxymethyl cellulose with functional groups $-COOH$ and $-OH$ showed more swelling in basic medium, and tertiary amine starch ether ($>N$, $-OH$ groups) shows acidic swelling (Kocak et al., 2017) (Figure 5.8).

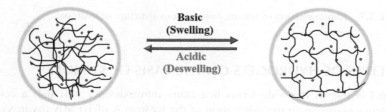

FIGURE 5.8 Swelling–deswelling response of hydrogel with variation of pH from basic to acidic.

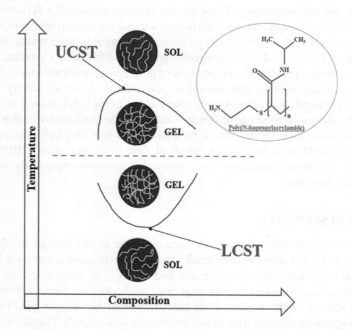

FIGURE 5.9 Temperature-sensitive hydrogels representing LCST and UCST behaviour.

5.10   TEMPERATURE RESPONSIVE

Temperature-sensitive or thermoresponsive hydrogels showed a radical alteration in their properties with respect to temperature change. The mechanical characteristics of the hydrogels and biological material altered when exposed to various temperatures. Due to the properties like sol–gel conversion and LCST (lower critical solution temperature), PNIPAAM (poly(N-iso-propyl acrylamide), poly AAm (polyacrylamide), and PDEAm poly(N,N-diethyl acrylamide) were extensively used for temperature-sensitive biomedical applications (Kano & Kokufuta, 2009). PNIPAAM (poly(N-iso-propyl acrylamide)) has gained a lot of attraction due to its excellent thermal response with the sharp temperature change and hence have numerous applications in the biomedical field as such or with slight modifications or along with bio-based backbones (Cheaburu-Yilmaz et al., 2018). The temperature at which the polymers or hydrogel converts from sol to gel or from soluble state to insoluble is known as cloud point (CP). The two important behaviours of the temperature-responsive hydrogels are LCST (lower critical solution temperature) and UCST (upper critical solution temperature), both addresses the opposite behaviour of the hydrogels (Frazar et al., 2020).

5.11   IONIC CONCENTRATION RESPONSIVE

Ionic strength is one of the easily controllable parameters or stimuli. With increase or decrease in ionic strength, hydrogel activity (swelling) is changed (García, 2018). Hydrogels consisting of cationic and ionic functionalities are more sensitive to the change in concentration. Variation in ionic concentration/salt concentration may lead to swelling–deswelling, micelles formation, solubility change, etc. (Chen et al., 2013).

5.11.1   BIO-MOLECULES

Similar to the above-explained stimuli, many biomolecules can act as a stimulus for the activity of hydrogels, such as enzymes, peptides, and glucose. Since hydrogels show different behaviours in the presence and absence of a biomolecule or increased or decreased concentration of a biomolecule, biosensors can be designed. B. Law et al. in 2007 worked on peptide-based hydrogels respondents to protease stimulus for protease-mediated drug delivery applications (Law et al., 2006). Kairali Podual and co-workers synthesised cationic copolymeric hydrogels of poly(ethylene glycol) responsive to enzymes such as catalase. The presence of the catalase enzyme was thought to increase the swelling significantly (Podual et al., 2000). Volkan Yesilyurt et al. fabricated self-healing, injectable glucose-responsive hydrogels of poly(ethylene glycol) (PEG) and phenylboronic acid (PBA) (2016).

5.11.2   EMPLOYABILITY

Hydrogels due to their unique and smart behaviour find applications in all the horizons of the industry and the biomedical field. They are well known for their application in food, fitters, industry, separation, enrichment, sensing, etc. There is no sector which is devoid of their applications. Due to high water absorbance, they are primarily used in agriculture, forestry, water conservation, drought management, etc. Very commonly, they are used to prepare optical lenses, baby diapers, hygiene products, wound dressing, and bandages. In one view, the use of hydrogels seems to be quite commonest but these are actually complex interactions between hydrogels and the surrounding environment (Aswathy et al., 2020).

5.12 BIOMEDICAL APPLICATIONS

Water being the utmost important component of all the living organisms including human beings, water absorbents and retainer hydrogels were considered to have great potential for application in the biomedical industry (Sharma & Tiwari, 2020). Biodegradability and biocompatibility are the properties of hydrogels, which makes them more advantageous than other biomaterials. But some issues such as solubility, mechanical strength, and thermal stability limit the horizon of employability of these hydrogels. Therefore, a number of researchers are working on the modification of hydrogels according to desirable properties. Bio-based polymers however have gained advantages over the synthetics because of their availability from natural resources, biodegradability, zero-waste, etc. (Shewan & Stokes, 2013) (Figure 5.10).

5.13 SUSTAINABLE DRUG RELEASE

One of the significant goals in case of target-triggered drug deliverance is to absorb sufficiently, hold controllable, transport site specifically and to release sustainably at the required site (Ghasemiyeh & Mohammadi-Samani, 2019). The idea of the target-oriented drug delivery originated in 1975 so as to minimise premature drug metabolism as well as drug toxicity and to maximise drug efficiency in the body (Narayanaswamy & Torchilin, 2019). Hydrogels are used as drug delivery devices for a longer time with a number of amendments. When the hydrogels are immersed

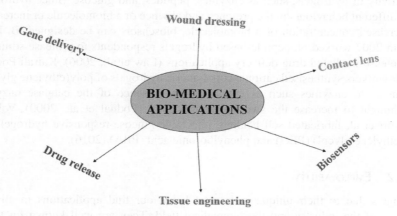

FIGURE 5.10 Diverse applications of the hydrogels in the biomedical field.

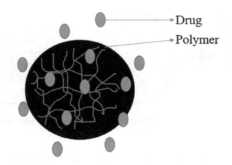

FIGURE 5.11   Drug–polymer interactions.

into an aqueous medium containing the drug, the drug enters into hydrogel through the pores along with water (Li & Mooney, 2016). Maximum drug upload into the hydrogel system can be achieved using the equilibration technique. Drug release in the body can be done by a number of ways. One way is stimulus responsive. It means to release the drug at the targeted position in a controlled and sustainable manner (Gupta et al., 2002). M. Prabaharan et al. covalently crosslink chitosan for controlled drug release (Prabaharan & Mano, 2005). The stimulus in the body (can be pH, temperature, glucose, etc.) triggers the hydrogel to deswell and release the content. Drug diffusion through the hydrogels can be affected by environmental factors such as temperature, pH, light, and pressure (Hoare & Kohane, 2008). Numerous models were applied to drug release to understand the mechanism by which the drug diffuses out of the system. Nadia Halib and his fellow researchers synthesised bacterial cellulose and acrylic acid-based hydrogels for the sustainable release of bovine serum albumin (Mohd Amin et al., 2012) (Figure 5.11).

5.14   GENE DELIVERY

Hydrogels are highly valuable biomaterials for human beings. Genome (RNA/DNA) delivery is the incorporation of genes into the hydrogels and transporting them to the desirable position. The literature survey showed that hydrogels with tissue-specific functionalities were used for sustainable, controlled, and localised genome delivery so that cells adapt it (Carballo-Pedrares et al., 2020). The genome delivery can be done by viral and non-viral methods. Clinically, the non-viral method of gene delivery is more accepted due to better biosafety, zero risk of mutagenesis, comparatively lowest risk of adverse immune reactions, higher transfection efficiency, etc. (Cao et al., 2019). Yan Li and co-workers worked with PEG (poly ethylene glycol)-based nano hydrogels for the gene deliverance in the mesenchymal stem cells of humans (Li et al., 2012). Di Chuan et al. had published a review article on gene delivery using biopolymer chitosan, and they have discussed various methods for improving the properties of chitosan, making it a suitable candidate for gene delivery (2019) (Figure 5.12).

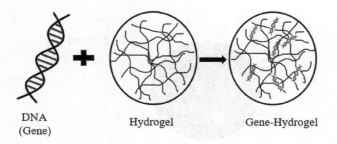

DNA
(Gene)   Hydrogel   Gene-Hydrogel

FIGURE 5.12   Pictorial representation of gene–hydrogel interaction.

5.15   TISSUE ENGINEERING

Tissue engineering is the branch which combines the principles of science and engineering to repair the damaged tissues/organs (Nguyen & West, 2002). Development of biomaterials which are biocompatible with the body cells helps to promote cellular regeneration and production of new tissues (Vinatier et al., 2006). These materials are bioactive and fill the gap in the tissue scaffolds. Dugan, Gough, and Eichhorn used scaffolds of bacterial cellulose and nanowhiskers of for tissue regeneration (Dugan et al., 2013) (Figure 5.13).

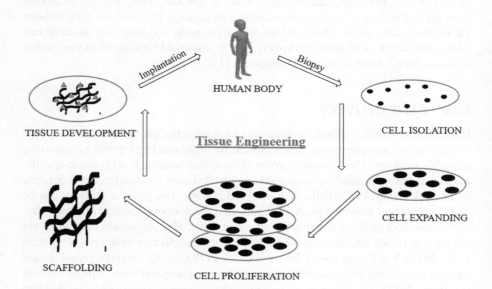

FIGURE 5.13   Employability of hydrogels in tissue regeneration.

5.16 BIOSENSING

There is an increasing demand for biosensors in the field of biotechnology to develop smart biomedical devices such as thermal screeners, eye scanners, biometric machines, blood pressure sensors, and glucose level sensors. The bioresponsive hydrogel candidates can be micro- and nano-patterned so as to prepare nanochips (Wolfbeis & Weidgans, 2006). A photonic crystal which is able to sense the increasing and decreasing glucose level in diabetic patients was prepared by Holtz and Asher. The hydrogel can act like a sensor because of its biocompatibility property along with sensing (Holtz et al., 1998).

5.17 WOUND DRESSING

The wound on the skin is breakage or defect occurring due to physicochemical or thermal damage to skin. It can be acute or chronic (Gobi et al., 2021). The human and the animal skin have a self-healing property by the generation of new cells and shattering of the damaged cells with due course of time. But certain gels and bandages will quicken the process of healing by speeding up the regeneration of new tissues (Wang et al., 2020). With growing technological uses in the biomedical fields, scientists are looking for biocompatible materials which are found to be quick and focussed for nursing wounds smartly. Hydrogel bandages when applied to the wounds not only act as physical barriers but also remove excess exudate to quicken the process of healing. Hydrogels encapsulate the bioactive materials and provide environment (moisture) for speedy healing. Mehrdad Kokabi and his fellow workers fabricated nanohydrogels based on clay for wound dressing and found that hydrogels were fulfilling all the vital necessities for the sensible wound dressing (Kokabi et al., 2007). Also, Sudhesh Kumar prepared novel antibacterial composite of chitin/nanosilver against *Staphylococcus aureus* and *Escherichia coli*. The composites also exhibit virtuous blood clotting capacity, proving suitable for wound dressing applications.

5.18 ANTIMICROBIAL APPLICATIONS

Antimicrobial agents eradicate or reduce the activity of the microorganisms onto wound or infection. Microorganisms include bacteria, virus, or fungi (Li et al., 2018). The body of the living organisms has a natural tendency to fight with the microbes but researchers with the advance of the hydrogels developed the engineered hydrogels acting as antimicrobial agents and work naturally as body's peptide defence mechanism. Antimicrobial hydrogels therefore are also used as fillers and wound dressing materials. The antibacterial properties of the hydrogels are also associated with wound dressing as well as wound healing applications (Hamedi et al., 2018). Ji Un Shin et al. fabricated nanoparticles of cellulose having Ag incorporated in it for antibacterial action of *E. coli*. The nanoparticles showed efficient activity against *E. coli* with pointedly less cytotoxicity (Shin et al., 2018). P. T. Sudheesh Kumar and co-workers prepared antibacterial bandages of chitin hydrogel/nanoZnO for wound dressing applications (Madhumathi et al., 2010).

5.19   FUTURE CHALLENGES AND SCOPE

Hydrogels have widespread applications from diaper linings to biosensors. The potential of hydrogels has been exposed with new concepts. In our daily life, we find many hydrogel materials in the surrounding such as floor, blond psyllium, starch, and gelatine. So, the concept is simple but the chemistry involved is quite complex. Various biomaterials can be synthesised according to the applicability, for example, for the adsorption of the metal ions from the water bodies, hydrogels with anionic functionalities are preferred or designed. For the delivery of the bioactive material, the desired functionalities were inculcated into the hydrogel backbones. As discussed above, various commercially important hydrogels are there in the market but yet there are some limitations on their usage because of their generation from synthetic chemicals. The hydrogels that are originated or fabricated from the natural materials or natural resources are still preferred over the synthetic ones because of the sustainability of the environment. Many researchers are producing hydrogels from waste materials such as agro waste, domestic waste, and forest waste which in turn is highly advantageous over others.

Along with the number of limitations working with the hydrogels, there are countless advantages too. In future, during fabrication of new hydrogel materials, one should fabricate these by keeping the applications in prior so that maximum efficient materials must be generated. Also, the work in the field of best hydrogel generation from biowaste or agro waste should be done at the industrial scale so that the synthetic hydrogels can be replaced with the biological ones. Also, by adopting these kinds of practices, we can promote the green technologies in the field of synthesis and applications. The concern must be to generate the materials with maximum efficiency and zero waste generation.

REFERENCES

Ajdary, R., Tardy, B. L., Mattos, B. D., Bai, L., & Rojas, O. J. (2021). Plant nanomaterials and inspiration from nature: Water interactions and hierarchically structured hydrogels. *Advanced Materials*, *33*(28). https://doi.org/10.1002/adma.202001085

Aswathy, S. H., Narendrakumar, U., & Manjubala, I. (2020). Commercial hydrogels for biomedical applications. *Heliyon*, *6*(4), e03719. https://doi.org/10.1016/j.heliyon.2020.e03719

Beckett, L. E., Lewis, J. T., Tonge, T. K., & Korley, L. S. T. J. (2020). Enhancement of the mechanical properties of hydrogels with continuous fibrous reinforcement. *ACS Biomaterials Science and Engineering*, *6*(10), 5453–5473. https://doi.org/10.1021/acsbiomaterials.0c00911

Best, J. P., Neubauer, M. P., Javed, S., Dam, H. H., Fery, A., & Caruso, F. (2013). Mechanics of pH-responsive hydrogel capsules. *Langmuir*, *29*(31), 9814–9823. https://doi.org/10.1021/la402111v

Can, V., Abdurrahmanoglu, S., & Okay, O. (2007). Unusual swelling behavior of polymer-clay nanocomposite hydrogels. *Polymer*, *48*(17), 5016–5023. https://doi.org/10.1016/j.polymer.2007.06.066

Cao, Y., Tan, Y. F., Wong, Y. S., Liew, M. W. J., & Venkatraman, S. (2019). Recent advances in chitosan-based carriers for gene delivery. *Marine Drugs*, *17*(6). https://doi.org/10.3390/md17060381

Carballo-Pedrares, N., Fuentes-Boquete, I., Díaz-Prado, S., & Rey-Rico, A. (2020). Hydrogel-based localized nonviral gene delivery in regenerative medicine approaches—an overview. *Pharmaceutics*, *12*(8), 1–21. https://doi.org/10.3390/pharmaceutics12080752

Chai, Q., Jiao, Y., & Yu, X. (2017). Hydrogels for biomedical applications: Their characteristics and the mechanisms behind them. *Gels*, *3*(1). https://doi.org/10.3390/gels3010006

Cheaburu-Yilmaz, C. N., Vasile, C., Ciocoiu, O.-N., & Staikos, G. (2018). Sodium Alginate Grafted with Poly(N -isopropylacrylamide). In *Temperature-Responsive Polymers* (Issue February 2019). https://doi.org/10.1002/9781119157830.ch5

Chen, J., Park, H., & Park, K. (1999). Synthesis of superporous hydrogels: Hydrogels with fast swelling and superabsorbent properties. *Journal of Biomedical Materials Research*, *44*(1), 53–62. https://doi.org/10.1002/(SICI)1097-4636(199901)44:1<53::AID-JBM6>3.0.CO;2-W

Chen, Y. C., Xie, R., & Chu, L. Y. (2013). Stimuli-responsive gating membranes responding to temperature, pH, salt concentration and anion species. *Journal of Membrane Science*, *442*, 206–215. https://doi.org/10.1016/j.memsci.2013.04.041

Chern, J. M., Lee, W. F., & Hsieh, M. Y. (2004). Preparation and swelling characterization of poly(n-isopropylacrylamide)- based porous hydrogels. *Journal of Applied Polymer Science*, *92*(6), 3651–3658. https://doi.org/10.1002/app.20301

Chuan, D., Jin, T., Fan, R., Zhou, L., & Guo, G. (2019). Chitosan for gene delivery: Methods for improvement and applications. *Advances in Colloid and Interface Science*, *268*, 25–38. https://doi.org/10.1016/j.cis.2019.03.007

Dalei, G., Das, S., & Das, S. P. (2019). Non-thermal plasma assisted surface nano-textured carboxymethyl guar gum/chitosan hydrogels for biomedical applications. *RSC Advances*, *9*(3), 1705–1716. https://doi.org/10.1039/C8RA09161G

Das, D., Das, R., Mandal, J., Ghosh, A., & Pal, S. (2014). Dextrin crosslinked with poly(lactic acid): A novel hydrogel for controlled drug release application. *Journal of Applied Polymer Science*, *131*(7), 1–12. https://doi.org/10.1002/app.40039

De, K. S., Aluru, N. R., Johnson, B., Crone, W. C., Beebe, D. J., & Moore, J. (2002). Equilibrium swelling and kinetics of pH-responsive hydrogels: Models, experiments, and simulations. *Journal of Microelectromechanical Systems*, *11*(5), 544–555. https://doi.org/10.1109/JMEMS.2002.803281

De las Heras Alarcón, C., Pennadam, S., & Alexander, C. (2005). Stimuli responsive polymers for biomedical applications. *Chemical Society Reviews*, *34*(3), 276–285. https://doi.org/10.1039/b406727d

Dolbow, J., Fried, E., & Ji, H. (n.d.). Chemically-induced swelling of hydrogels. *Journal of the Mechanics and Physics of Solids*, *52*(1), 51–84.

Du, H., Liu, W., Zhang, M., Si, C., Zhang, X., & Li, B. (2019). Cellulose nanocrystals and cellulose nanofibrils based hydrogels for biomedical applications. *Carbohydrate Polymers*, *209*(November 2018), 130–144. https://doi.org/10.1016/j.carbpol.2019.01.020

Dugan, J. M., Gough, J. E., & Eichhorn, S. J. (2013). Bacterial cellulose scaffolds and cellulose nanowhiskers for tissue engineering. *Nanomedicine*, *8*(2), 287–298. https://doi.org/10.2217/nnm.12.211

Fertahi, S., Ilsouk, M., Zeroual, Y., Oukarroum, A., & Barakat, A. (2021). Recent trends in organic coating based on biopolymers and biomass for controlled and slow release fertilizers. *Journal of Controlled Release*, *330*(January 2021), 341–361. https://doi.org/10.1016/j.jconrel.2020.12.026

Frazar, E. M., Shah, R. A., Dziubla, T. D., & Hilt, J. Z. (2020). Multifunctional temperature-responsive polymers as advanced biomaterials and beyond. *Journal of Applied Polymer Science*, *137*(25), 1–14. https://doi.org/10.1002/app.48770

García, M. C. (2018). Ionic-Strength-Responsive Polymers for Drug Delivery Applications. In *Stimuli Responsive Polymeric Nanocarriers for Drug Delivery Applications: Volume 2: Advanced Nanocarriers for Therapeutics*. Elsevier Ltd. https://doi.org/10.1016/B978-0-08-101995-5.00014-3

Garrison, T. F., Murawski, A., & Quirino, R. L. (2016). Bio-based polymers with potential for biodegradability. *Polymers, 8*(7), 1–22. https://doi.org/10.3390/polym8070262

Ghasemiyeh, P., & Mohammadi-Samani, S. (2019). Hydrogels as drug delivery systems; pros and cons. *Trends in Pharmaceutical Sciences, 5*(1), 7–24. https://doi.org/10.30476/TIPS.2019.81604.1002

Gobi, R., Ravichandiran, P., Babu, R. S., & Yoo, D. J. (2021). Biopolymer and synthetic polymer-based nanocomposites in wound dressing applications: A review. *Polymers, 13*(12). https://doi.org/10.3390/polym13121962

Gupta, P., Vermani, K., & Garg, S. (2002). Hydrogels: From controlled release to pH-responsive drug delivery. *Drug Discovery Today, 7*(10), 569–579. https://doi.org/10.1016/S1359-6446(02)02255-9

Hamedi, H., Moradi, S., Hudson, S. M., & Tonelli, A. E. (2018). Chitosan based hydrogels and their applications for drug delivery in wound dressings: A review. *Carbohydrate Polymers, 199*(March), 445–460. https://doi.org/10.1016/j.carbpol.2018.06.114

Hennink, W. E., & van Nostrum, C. F. (2012). Novel crosslinking methods to design hydrogels. *Advanced Drug Delivery Reviews, 64*(SUPPL.), 223–236. https://doi.org/10.1016/j.addr.2012.09.009

Hoare, T. R., & Kohane, D. S. (2008). Hydrogels in drug delivery: Progress and challenges. *Polymer, 49*(8), 1993–2007. https://doi.org/10.1016/j.polymer.2008.01.027

Holtz, J. H., Holtz, J. S. W., Munro, C. H., & Asher, S. A. (1998). Arrays: Novel chemical sensor materials. *Changes, 70*(4), 780–791.

Hongbo, T., Yanping, L., Min, S., & Xiguang, W. (2012). Preparation and property of crosslinking guar gum. *Polymer Journal, 44*(3), 211–216. https://doi.org/10.1038/pj.2011.117

Huang, R., Zheng, S., Liu, Z., & Ng, T. Y. (2020). Recent advances of the constitutive models of smart materials - Hydrogels and shape memory polymers. *International Journal of Applied Mechanics, 12*(2). https://doi.org/10.1142/S1758825120500143

Kano, M., & Kokufuta, E. (2009). On the temperature-responsive polymers and gels based on N-propylacrylamides and N-propylmethacrylamides. *Langmuir, 25*(15), 8649–8655. https://doi.org/10.1021/la804286j

Kawaguchi, H., Ogino, C., & Kondo, A. (2017). Microbial conversion of biomass into bio-based polymers. *Bioresource Technology, 245*, 1664–1673. https://doi.org/10.1016/j.biortech.2017.06.135

Kocak, G., Tuncer, C., & Bütün, V. (2017). PH-responsive polymers. *Polymer Chemistry, 8*(1), 144–176. https://doi.org/10.1039/c6py01872f

Kokabi, M., Sirousazar, M., & Hassan, Z. M. (2007). PVA-clay nanocomposite hydrogels for wound dressing. *European Polymer Journal, 43*(3), 773–781. https://doi.org/10.1016/j.eurpolymj.2006.11.030

Kono, H., & Fujita, S. (2012). Biodegradable superabsorbent hydrogels derived from cellulose by esterification crosslinking with 1,2,3,4-butanetetracarboxylic dianhydride. *Carbohydrate Polymers, 87*(4), 2582–2588. https://doi.org/10.1016/j.carbpol.2011.11.045

Kornblum, S., Hasbroucq, T., & Osman, A. (1990). Dimensional overlap: Cognitive basis for stimulus-response compatibility-a model and taxonomy. *Psychological Review, 97*(2), 253–270. https://doi.org/10.1037/0033-295X.97.2.253

Law, B., Weissleder, R., & Tung, C. H. (2006). Peptide-based biomaterials for protease-enhanced drug delivery. *Biomacromolecules, 7*(4), 1261–1265. https://doi.org/10.1021/bm050920f

Li, H., Yuan, Z., Lam, K. Y., Lee, H. P., Chen, J., Hanes, J., & Fu, J. (2004). Model development and numerical simulation of electric-stimulus-responsive hydrogels subject to an externally applied electric field. *Biosensors and Bioelectronics*, *19*(9), 1097–1107. https://doi.org/10.1016/j.bios.2003.10.004

Li, J., & Mooney, D. J. (2016). Designing hydrogels for controlled drug delivery. *Nature Reviews Materials*, *1*(12), 1–18. https://doi.org/10.1038/natrevmats.2016.71

Li, S., Dong, S., Xu, W., Tu, S., Yan, L., Zhao, C., Ding, J., & Chen, X. (2018). Antibacterial hydrogels. *Advanced Science*, *5*(5). https://doi.org/10.1002/advs.201700527

Li, Y., Yang, C., Khan, M., Liu, S., Hedrick, J. L., Yang, Y. Y., & Ee, P. L. R. (2012). Nanostructured PEG-based hydrogels with tunable physical properties for gene delivery to human mesenchymal stem cells. *Biomaterials*, *33*(27), 6533–6541. https://doi.org/10.1016/j.biomaterials.2012.05.043

Madhumathi, K., Sudheesh Kumar, P. T., Abhilash, S., Sreeja, V., Tamura, H., Manzoor, K., Nair, S. V., & Jayakumar, R. (2010). Development of novel chitin/nanosilver composite scaffolds for wound dressing applications. *Journal of Materials Science: Materials in Medicine*, *21*(2), 807–813. https://doi.org/10.1007/s10856-009-3877-z

Maitra, J., & Shukla, V. K. (2014). Cross-linking in hydrogels - A review. *American Journal of Polymer Science*, *4*(2), 25–31. https://doi.org/10.5923/j.ajps.20140402.01

Manuscript, A. (2019). *pt*.

Mohd Amin, M. C. I., Ahmad, N., Halib, N., & Ahmad, I. (2012). Synthesis and characterization of thermo- and pH-responsive bacterial cellulose/acrylic acid hydrogels for drug delivery. *Carbohydrate Polymers*, *88*(2), 465–473. https://doi.org/10.1016/j.carbpol.2011.12.022

Mondal, S., Das, S., & Nandi, A. K. (2020). A review on recent advances in polymer and peptide hydrogels. *Soft Matter*, *16*(6), 1404–1454. https://doi.org/10.1039/c9sm02127b

Muñoz-Bonilla, A., Echeverria, C., Sonseca, Á., Arrieta, M. P., & Fernández-García, M. (2019). Bio-based polymers with antimicrobial properties towards sustainable development. *Materials*, *12*(4). https://doi.org/10.3390/ma12040641

Nakajima, H., Dijkstra, P., & Loos, K. (2017). The recent developments in biobased polymers toward general and engineering applications: Polymers that are upgraded from biodegradable polymers, analogous to petroleum-derived polymers, and newly developed. *Polymers*, *9*(10), 1–26. https://doi.org/10.3390/polym9100523

Narayanaswamy, R., & Torchilin, V. P. (2019). Hydrogels and their applications in targeted drug delivery. *Molecules*, *24*(3). https://doi.org/10.3390/molecules24030603

Nath, N., & Chilkoti, A. (2002). Creating "smart" surfaces using stimuli responsive polymers. *Advanced Materials*, *14*(17), 1243–1247. https://doi.org/10.1002/1521-4095(20020903)14:17<1243::AID-ADMA1243>3.0.CO;2-M

Nguyen, K. T., & West, J. L. (2002). Photopolymerizable hydrogels for tissue engineering applications. *Biomaterials*, *23*(22), 4307–4314. https://doi.org/10.1016/S0142-9612(02)00175-8

Peppas, N. A., Hilt, J. Z., Khademhosseini, A., & Langer, R. (2006). Hydrogels in biology and medicine: From molecular principles to bionanotechnology. *Advanced Materials*, *18*(11), 1345–1360. https://doi.org/10.1002/adma.200501612

Pimenta, A. F. R., Serro, A. P., Colaço, R., & Chauhan, A. (2019). Optimization of intraocular lens hydrogels for dual drug release: Experimentation and modelling. *European Journal of Pharmaceutics and Biopharmaceutics*, *141*, 51–57. https://doi.org/10.1016/j.ejpb.2019.05.016

Podual, K., Doyle, F. J., & Peppas, N. A. (2000). Glucose-sensitivity of glucose oxidase-containing cationic copolymer hydrogels having poly(ethylene glycol) grafts. *Journal of Controlled Release*, *67*(1), 9–17. https://doi.org/10.1016/S0168-3659(00)00195-4

Pourjavadi, A., Heydarpour, R., & Tehrani, Z. M. (2021). Multi-stimuli-responsive hydrogels and their medical applications. *New Journal of Chemistry*, *45*(35), 15705–15717. https://doi.org/10.1039/d1nj02260a

Prabaharan, M., & Mano, J. F. (2005). Chitosan-based particles as controlled drug delivery systems. *Drug Delivery: Journal of Delivery and Targeting of Therapeutic Agents*, *12*(1), 41–57. https://doi.org/10.1080/10717540590889781

Sajid, M. A., Shahzad, S. A., Hussain, F., Skene, W. G., Khan, Z. A., & Yar, M. (2018). Synthetic modifications of chitin and chitosan as multipurpose biopolymers: A review. *Synthetic Communications*, *48*(15), 1893–1908. https://doi.org/10.1080/00397911.2018.1465096

Sharma, S., & Tiwari, S. (2020). A review on biomacromolecular hydrogel classification and its applications. *International Journal of Biological Macromolecules*, *162*, 737–747. https://doi.org/10.1016/j.ijbiomac.2020.06.110

Sheth, R. D., Madan, B., Chen, W., & Cramer, S. M. (2013). High-throughput screening for the development of a monoclonal antibody affinity precipitation step using ELP-z stimuli responsive biopolymers. *Biotechnology and Bioengineering*, *110*(10), 2664–2676. https://doi.org/10.1002/bit.24945

Shewan, H. M., & Stokes, J. R. (2013). Review of techniques to manufacture micro-hydrogel particles for the food industry and their applications. *Journal of Food Engineering*, *119*(4), 781–792. https://doi.org/10.1016/j.jfoodeng.2013.06.046

Shin, J. U., Gwon, J., Lee, S. Y., & Yoo, H. S. (2018). Silver-incorporated nanocellulose fibers for antibacterial hydrogels. *ACS Omega*, *3*(11), 16150–16157. https://doi.org/10.1021/acsomega.8b02180

Silva, D. M., Nunes, C., Pereira, I., Moreira, A. S. P., Domingues, M. R. M., Coimbra, M. A., & Gama, F. M. (2014). Structural analysis of dextrins and characterization of dextrin-based biomedical hydrogels. *Carbohydrate Polymers*, *114*, 458–466. https://doi.org/10.1016/j.carbpol.2014.08.009

Van Vlierberghe, S., Cnudde, V., Dubruel, P., Masschaele, B., Cosijns, A., De Paepe, I., Jacobs, P. J. S., Van Hoorebeke, L., Remon, J. P., & Schacht, E. (2007). Porous gelatin hydrogels: 1. Cryogenic formation and structure analysis. *Biomacromolecules*, *8*(2), 331–337. https://doi.org/10.1021/bm060684o

Velusamy, S., Roy, A., Sundaram, S., & Kumar Mallick, T. (2021). A review on heavy metal ions and containing dyes removal through graphene oxide-based adsorption strategies for textile wastewater treatment. *Chemical Record*, *21*(7), 1570–1610. https://doi.org/10.1002/tcr.202000153

Vinatier, C., Guicheux, J., Daculsi, G., Layrolle, P., & Weiss, P. (2006). Cartilage and bone tissue engineering using hydrogels. *Bio-Medical Materials and Engineering*, *16*(4 SUPPL.), S107–S113.

Wang, L., & Stegemann, J. P. (2011). Glyoxal crosslinking of cell-seeded chitosan/collagen hydrogels for bone regeneration. *Acta Biomaterialia*, *7*(6), 2410–2417. https://doi.org/10.1016/j.actbio.2011.02.029

Wang, Y., Huang, W., Huang, W., Wang, Y., Mu, X., Ling, S., Yu, H., Chen, W., Guo, C., Watson, M. C., Yu, Y., Black, L. D., Li, M., Omenetto, F. G., Li, C., & Kaplan, D. L. (2020). Stimuli-responsive composite biopolymer actuators with selective spatial deformation behavior. *Proceedings of the National Academy of Sciences of the United States of America*, *117*(25), 14602–14608. https://doi.org/10.1073/pnas.2002996117

Wilson, A. N., & Guiseppi-Elie, A. (2013). Bioresponsive hydrogels. *Advanced Healthcare Materials*, *2*(4), 520–532. https://doi.org/10.1002/adhm.201200332

Wolfbeis, O. S., & Weidgans, B. M. (2006). Fiber optic chemical sensors and biosensors: a view back. In *Optical chemical sensors* (pp. 17–44). Springer, Dordrecht.

Yesilyurt, V., Webber, M. J., Appel, E. A., Godwin, C., Langer, R., & Anderson, D. G. (2016). Injectable self-healing glucose-responsive hydrogels with pH-regulated mechanical properties. *Advanced Materials*, *28*(1), 86–91. https://doi.org/10.1002/adma.201502902

Zhang, Y., Wang, Q. S., Yan, K., Qi, Y., Wang, G. F., & Cui, Y. L. (2016). Preparation, characterization, and evaluation of genipin crosslinked chitosan/gelatin three-dimensional scaffolds for liver tissue engineering applications. *Journal of Biomedical Materials Research - Part A*, *104*(8), 1863–1870. https://doi.org/10.1002/jbm.a.35717

Zhao, Y., Zhu, Z. S., Guan, J., & Wu, S. J. (2021). Processing, mechanical properties and bio-applications of silk fibroin-based high-strength hydrogels. *Acta Biomaterialia*, *125*, 57–71. https://doi.org/10.1016/j.actbio.2021.02.018

Zhu, K., Shi, S., Cao, Y., Lu, A., Hu, J., & Zhang, L. (2019). Robust chitin films with good biocompatibility and breathable properties. *Carbohydrate Polymers*, *212*(February), 361–367. https://doi.org/10.1016/j.carbpol.2019.02.054

6 Scaffold-Based Tissue Engineering for Craniofacial Deformities

Jasmine Nindra and Mona Prabhakar
Shree Guru Gobind Singh Tricentenary University,
Gurugram, India

CONTENTS

DOI: 10.1201/9781003266464-6

6.1 INTRODUCTION

Craniofacial deformities include a large fraction of human bone defects, requiring interdisciplinary intervention by various specialists for surgical, nutritional, dental, speech, medical, and behavioral therapies that impose a substantial economic and societal burden. These deformities occur secondary to trauma, recession of bone due to tumor and cyst, or congenital defects and may affect both hard and soft tissues. The most prevalent congenital craniofacial defects are cleft lip and palate, which are caused by disturbance during embryonic development of hard and soft tissues of the oral cavity, resulting in non-fusion of the two palatine shelves (Martín-Del-Campo et al., 2019). Extensive surgical intervention using autologous bone grafting techniques is required for correction of these defects, which involves prolonged healing time at the donor and correction site, risk of infection, postoperative pain, and risk of graft failure. The child requiring craniofacial defect correction, along with their family, has to go through huge emotional as well as financial trauma that ultimately places financial burden on the healthcare system[3]. Due to the known challenges of extensive surgical interventions, the reconstructive surgeons have turned to scientists and engineers for alternate approaches of correction of these defects, in order to reduce severity of debilitating effects associated with surgery. It is quite impossible to completely eliminate surgery; however, the creation of high-end technologies improves surgical outcome by reducing the required number of surgical procedures. One such alternative technology to conventional autogenous bone graft techniques, which involves using scaffolds, stem cells, and signaling molecules to achieve therapeutic goals, is known as tissue engineering. This synergistic triad of functional biomaterials, bioactive molecules, and recruited stem cells improves life conditions of an affected individual by enhancing the self-repairment mechanism of affected tissues. Due to complexity of the maxillofacial region, it is necessary for the reconstruction procedures to maintain anatomic uniformity and appearance along with restoring tissue functions (Bhumiratana & Vunjak-Novakovic, 2012). Thus, to enhance mechanical, functional, and regenerative properties of grafts, tissue-engineered constructs are fabricated that imitate the original structure of matrix (scaffold), cells and bioactive molecules (Farré-Guasch et al., 2015). For adequate dimensional and structural properties, scaffold can be designed using additive manufacturing (AM) or 3D printing technologies customized according to individual patient needs. A recent advancement to AM is the 3D bioprinting technique, in which osteoinductive growth factors such as stem cells and adipose tissue cells are incorporated within or onto the printed scaffold[6].

6.2 CRANIOFACIAL DEFORMITIES

Craniofacial deformities are a diverse group of malformations in which the normal anatomy of skull, jaws, and adjacent soft tissues is severely impaired. These deformities may be present at the time of birth known as congenital anomalies or inflicted later by means of a traumatic event through sports or various other accidents known as acquired deformities (Elsten et al., 2020). One of the most prevalent congenital craniofacial defects is cleft lip with and without palate that affects 1 in 700 live births and also comprises the second most common congenital anomaly after club foot (Witt et al., 1999). Cleft lip and palate appear as a split in the lip or palate, which is identifiable at the time of birth. It occurs due to atypical embryological development resulting from non-fusion of the two maxillary processes, which form majority of palate, during 8–12 weeks of intra-uterine life. This interruption in palatogenesis is characterized by a defect in the oronasal separation, which requires surgical intervention to close the defect using various grafting techniques (Witt et al., 1999). This loss in continuity of craniofacial tissue occurring secondary to congenital or acquired conditions is associated with significant function and aesthetic and psychological affliction. Reconstruction of craniofacial deformities is a challenging procedure due to the complex nature of the surrounding anatomical structures such as sensory organs, facial skeletal features, cartilage, and blood vessels (Oliver et al., 2021). The gold standard reconstructive treatment of these craniofacial bony defects is bone grafting using autografts and allografts. However, these are associated with drawbacks of donor site morbidity in case of autografts and the potential immunogenic rejection in case of allografts (Tian et al., 2018).

6.3 TISSUE ENGINEERING

Tissue engineering technique restores or replaces a damaged or diseased tissue through a synergistic triad that includes (Figure 6.1):

1. Design of scaffold matrices
2. Selection and manipulation of stem cells
3. Use of suitable biologic molecular signals

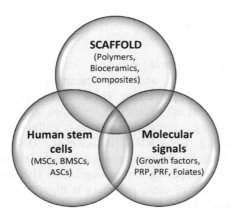

FIGURE 6.1 Synergistic triad of tissue engineering for regenerative bone therapy in cranio-facial deformities.

6.3.1 SCAFFOLD MATRICES

Scaffold matrix is fabricated to provide a solid framework for the growth and differentiation of seeded cells, which allows cell attachment and migration to facilitate development of a desired tissue.

Ideal properties of a scaffold for tissue engineering are (Salgado et al., 2004):

1. Biocompatibility: refers to the property of scaffolds by which it should support a normal cellular activity and get incorporated into the host tissue without any adverse immune response.
2. Porosity: scaffold should be porous enough with high interconnectivity to facilitate cell migration, proliferation, and growth along with nutrient diffusion and removal of metabolic waste.
3. Pore size: the diameter of the pore determines the amount of cell penetration, bone matrix production, and neovascularization.
4. Biofunctionality: the scaffold should meet the ideal functional requirements of the replaced tissue for which it is designed.
5. Osteoinductivity: the scaffold should act as mold to facilitate osteogenic differentiation to form the desired anatomical structure.
6. Biodegradability: the scaffold should provide a framework with adequate strength to allow growth of cells that must eventually degrade once the tissue regeneration procedure is complete. Materials used for scaffold fabrication can be divided into four categories (Ghassemi et al., 2018), which are discussed in the subsequent sections.

6.3.2 POLYMERIC SCAFFOLDS

Due to the design flexibility offered by polymeric materials, the physiochemical characteristics of scaffolds designed can be controlled, such as pore size, interconnection, porosity, solubility, allergic response, biocompatibility, and enzymatic reactions (Fuchs et al., 2001)

Commonly used polymers can be:

1. Natural polymers: Natural polymers show great biocompatibility, controlled biodegradation, and osteoinductive properties; however, they have poor mechanical properties. These are composed of extracellular biomaterials classified as proteins (e.g., collagen, gelatin, fibrinogen, etc.); polysaccharides (e.g., glycosaminoglycans, cellulose, dextran, chitin); polynucleotides (e.g., DNA, RNA) (Ratner et al., 2004)
2. Synthetic polymers: Synthetic polymers are easily fabricated into different shapes and are biocompatible and biodegradable (Ishaug et al., 1994).The compressive strength of these polymers is high; however, they undergo rapid in vivo degradation resulting in loss of strength and adverse tissue reactions. They consist of aliphatic polyesters such as polylactic-acid (PLA), polyethylene glycol (PEG), polycaprolactone (PCL), polyglycolic-acid (PGA), polyvinyl alcohol (PVA), poly propylene fumarate (PPF), polyacrylic acid (PAA), and polyurethane (PUR & PU) (Wang, 2003).

6.3.3   CERAMIC SCAFFOLDS

Bioceramics mimic the bone tissue, as they provide higher osteoblastic adherence and proliferation (Ducheyne & Qiu, 1999). The mechanical strength of ceramics is superior than polymers; however, in terms of tensile and torsion strength, they are still inferior to natural bone. They show both osteoconductive and osteoinductive properties. Common types of calcium phosphate ceramics used are tricalcium phosphate, hydroxyapatite, and their combinations as biphasic and amorphous calcium phosphates (Dorozhkin & Epple, 2002). Studies have shown that doping tricalcium phosphate scaffolds with ZnO (0.25%) and SiO_2 (0.5%) bring about a 2.5-fold increase in compressive strength and 92% increase in cell viability (Fielding et al., 2012).

6.3.4   COMPOSITE SCAFFOLDS

Two or more different materials like ceramic, metal, and polymers are combined to form a bioactive composite scaffold. Composites of polymers/ceramics provide excellent mechanical properties with improved osteoconductivity and controlled degradation for tissue engineering applications. Significantly used composite material in bone tissue engineering includes natural bone bio ceramics such as calcium phosphate, hydroxyapatite, and tricalcium phosphate with polylactic acid, collagen, gelatin, and chitosan (Zhang & Zhang, 2001). High-density polyethylene (HDPE) and poly(L-lactide-co-glycolide acid) (PLGA) are reinforced with hydroxyapatite to match the bone properties and matrices for bone mineralization and cell differentiation.

6.3.5   METALLIC SCAFFOLDS

Metal scaffolds act as permanent implants rather than scaffolds, as they are unrecognizable by biological factors. Metallic scaffolds are biocompatible and osteoconductive; however, they inhibit markers of bone formation, stimulation of bone loss, poorly osseointegrate with surrounding bone, and may release toxic ions due to corrosion.

For bone scaffolds, iron- and magnesium-based metals have been widely used such as Mg-Ca, Mg-RE (rare earth) alloys, pure Fe, and Fe-Mn alloys (Hermawan et al., 2008). Magnesium-based metals show suitable mechanical properties, high biodegradability, bio-resorbability, and bone cell activation support (Staiger et al., 2006; Witte et al., 2005). Modulus of elasticity of Mg and its alloys is closer to bone compared with other metals. Modifications in pore size to match its strength to that of bone can be done; however, high porosity can reduce corrosion resistance of magnesium. Thus, use of magnesium in medical science is limited due to its high corrosion and toxins.

Titanium porous scaffolds are used as bone replacement materials. They show high biocompatibility; however, they are non-biodegradable and do not integrate with biomolecules.[12] An alloy of titanium with nickel, that is, Nitinol (NiTi) show high biocompatibility and significant plasticity for bone scaffolding, but due allergic response and toxicity of Ni ions, its use is banned in America and Europe

(Assad et al., 2002). Tantalum (Ta) is another metal that can be used in bone tissue engineering as its elasticity is similar to that of bone.

6.3.6 METHODS OF DESIGNING A SCAFFOLD

Conventional methods: Involves a physicochemical method to ensure internal structures with pore size between 100 and 500 microns and porosities up to 90%. The disadvantage is randomly arranged trabeculae with physical properties that are difficult to control.

- Solvent casting/particle leaching
  The procedure involves selective leaching of water-soluble inorganic salts (KCl and NaCl) or organic compounds (gelatin), which are used as porogens. The polymer is first dissolved in high-volatile solvent and then the porogen, and the solution is casted into a mold. Once the solvent is evaporated and the porogens are dissolved to form a composite, a porous structure of scaffold is created.
- Thermally induced phase separation
  The process includes demixing of a uniform solution of polymer–solvent into polymer-lean and polymer-rich phases through quenching, thereby forming pores, once the solvent is evaporated.

Electrospinning method is a technique used to fabricate nanofiber polymers and composites. It consists of syringe pump, a collector, and high-voltage power source. The polymer droplet is held at syringe tip by surface tension which is counteracted by the electrostatic force caused by high-voltage source that erupts the electrically charged jet of polymer solution into a cone onto the collector that solidifies after evaporation of solvent. Electrospinning is widely used in fabrication of tissue scaffolds due its ability to generate micro- and nano-scale fibers and to process a wide range of polymers. Using this technique, scaffolds having well-interconnected uniform pores on micro-scale and a specific 3D structure on the macro-scale can be fabricated (Doshi & Reneker, 1995).

6.3.7 ADDITIVE MANUFACTURING/3D ACELLULAR PRINTING

The structural and dimensional requirements of complex external shapes and internal structures of craniofacial complex can be met by using AM techniques. These are automated and integrated with image techniques to produce customized scaffolds for individual patients. This method creates products by adding materials layer by layer to generate an arbitrary geometry. A computer-aided design software is used to design the 3D model geometry which is sent to a 3D printer to achieve the final product (Zhao et al., 2018).

AM technology is divided into the following categories.

a. Liquid-based processes:
  a. **Stereolithography**: a liquid photoreactive resin is added one layer at a time and cured using ultraviolet laser to manufacture the construct.

Three-dimensional models are created using this technique for diagnosis, planning a treatment, and simulation of the planned treatment procedure.

b. **Inkjet printing:** maxillofacial deformities are successfully treated with inkjet-printed, custom-made artificial bones of tricalcium phosphate.

c. **Jetted photopolymer printing:** the techniques used in inkjet printing and stereolithography are combined for fabricating high accuracy models which can be used for accurate preoperative planning and educating the patient by simulating the treatment plan.

b. Powder-based processes:

a. **Selective laser sintering:** powdered polymers or their composites are sintered using carbon dioxide laser to form solid 3D objects. The biodegradable polymers and bioactive ceramics can be combined by this method to successfully develop a completely biodegradable and osteoconductive nanocomposite scaffolds.

b. **Direct metal laser sintering:** layer-by-layer, metal and powder are fused using a focused laser beam into a localized region. Customized titanium metal meshes and reconstruction plates are manufactured using this method.

c. Solid-based processes:

a. **Fused deposition modeling (FDM):** a thermoplastic polymeric material is heated and pushed out in layers through a nozzle to fabricate a highly reproducible construct with a dimensional resolution of 0.2 mm and high neovascularization potential, making it suitable for fabricating large cell-scaffold structures.

Limitation of AM technologies is that it is difficult to integrate cells during fabrication of scaffold as the processing temperatures and pressures required during sintering are high, and contact with cytotoxic solvents (ethanol) makes the process parameters non-physiologic. The following section describes a promising alternative to conventional 3D printing, known as 3D bioprinting.

6.3.8   3D BIOPRINTING

This approach involves precise deposition of cells into 3D patterns using computer-aided software. The procedure involves incorporating cells, biologically active compounds, and extracellular matrix components within or onto a printed scaffold (Nyberg et al., 2017). Bioprinting method differs from that used for traditional printing since it is of utmost importance to maintain viability of cell during the printing process. Parameters of importance for 3D bioprinting include: cell positioning; bioink selection, and mechanical strength. The advantage of this approach is that cells are seeded into 3D-printed scaffold, by digitally designing deposition of cells layer-by-layer to precisely regulate the 3D distribution of cells. In contrast to traditional AM, bioinks used in bioprinting have lower mechanical strength than thermoplastic polymers. This can be overcome by integrating acellular and cellular bioprinting, for example, a 3D-printed muscle-tendon unit was created using extrusion bioprinting and FDM to print two thermoplastic polymers along with C2C12 and NIH/3T3 cells (Merceron et al., 2015).

A scaffold designed for craniofacial defect correction should maintain a balance between the degradation process and biomechanical properties to impart flexibility and stability that encourages de novo bone infiltration to totally restore the defect which is prompted by endogenous and exogenous biological cues such as growth factors, cells, and small molecules.

6.4 SELECTION AND MANIPULATION OF STEM CELLS

In bone tissue regeneration, use of mesenchymal stem cells (MSCs) for various cell-based therapies has been evaluated, as they have potential to differentiate into osteoblasts under certain conditions (Gładysz & Hozyasz, 2015). They are harvested using minimally invasive technique of needle aspiration from the donor site, in contrast to a more invasive harvesting procedure for autologous bone grafting. MSCs are known to modulate immune response and promote tissue regeneration and can differentiate into osteocytes, adipocytes, and chondrocytes. They are commonly extracted from bone marrow and adipose tissue and other organ types, including cartilage, umbilical cord, teeth, heart, kidney, and lung (Lam et al., 2020).

6.4.1 BONE MARROW–DERIVED STEM CELLS

Bone marrow–derived stem cells (BMSCs) are widely used in bone tissue engineering, as they show superior potential for osteogenic and chondrogenic differentiation.[9] A study comparing the traditional method of iliac crest bone grafting and the newer approach of using scaffold made of collagen seeded with autologous BMSCs, platelet-rich fibrin (PRF), and a nanohydroxyapatite for cleft palate repair, demonstrated improved healing and a reduced postoperative pain in the scaffold-based therapy (Al-Ahmady et al., 2018).

6.4.2 ADIPOSE-DERIVED STEM CELLS

Adipose-derived stem Cells (ADSCs) are a useful alternative to BMSCs and traditional bone grafts, as they have high stem cell to volume ratio, reduced sensitivity to aging, can be easily harvested and processed within a short time frame, and ability to undergo expansion and differentiation into various tissue types in vivo. It was observed to act as a preferable alternative to autografts, when undifferentiated ADSCs along with hydroxyapatite–tricalcium phosphate scaffolds were placed in maxillary alveolar cleft defects, as there is reduced operative time and lack of morbidity at donor site (Pourebrahim et al., 2013).

6.4.2.1 Tooth-Derived Stem Cells

There are five types of MSCs in dental population: (1) dental follicle progenitor stem cells; (2) stem cells from apical papilla; (3) periodontal ligament stem cells; (4)

dental pulp stem cells (DPSCs); and (5) stem cells from human exfoliated deciduous teeth (SHEDs). Among these, SHEDs can be easily isolated and extracted, show high proliferation ability, have capacity to undergo multilineage differentiation, and can modulate propagation of cytokine signals. However, a recent study comparing human BMSCs, SHEDs, and human-derived dental pulp stem cells concluded that there is no difference in the bone regenerative capacity between these cell populations (Nakajima et al., 2018).

6.5   USE OF BIOLOGIC SIGNALING MOLECULES

Biologic signaling molecules such as growth factors are naturally secreted signaling proteins, hormones, and peptides that bind to the receptors and transmit intracellular signals to regulate proliferation and differentiation of undifferentiated cells to increase or decrease specific cell populations. Tissue regeneration is enhanced using biochemical signals, in the form of growth factors or genes, that regulate cellular responses, and may lead to promotion or prevention of cell adhesion, proliferation, migration, and differentiation by up- or down-regulating the synthesis of proteins, other growth factors, or receptors.[9] Growth factors involved in regulating bone metabolism are:

6.5.1   BONE MORPHOGENETIC PROTEINS

Bone morphogenetic proteins (BMPs) have the potential to stimulate de novo osteogenesis that drives many processes in the formation of bone, starting from MSC migration to osteoblast differentiation (Carreira et al., 2014). Recombinant human BMP-2 and -7 are commercially available for use in cleft palate to improve local bone regeneration.

6.5.2   PLATELET-DERIVED GROWTH FACTORS

Isoforms of recombinant human platelet-derived growth factors (rhPDGFs), including AA, -BB, and -AB, are secreted from circulating platelets that act as potentiators in the morphogenesis of cranial and cardiac neural crest, skeleton, blood vessels, lung, and intestine. Purified rhPDGF therapy along with an osteoconductive matrix carrier (tricalcium phosphate) has been demonstrated to significantly fill the alveolar bone defectβ (Aichelmann-Reidy & Reynolds, 2008).

6.5.3   VASCULAR ENDOTHELIAL GROWTH FACTOR

Vascular endothelial growth factor (VEGF) promotes angiogenesis which is of critical importance for adequate healing at the site of bone grafts. VEGF is considered as a key regulator of vascular growth, required during skeletal development and postnatal bone repair for effective coupling of angiogenesis and osteogenesis (Zisch et al., 2003).

6.5.4 Fibroblast Growth Factors

Fibroblast growth factors (FGF-2) stimulate proliferation of periosteal cells, osteoprogenitors, and chondrogenitors, which results in bony callus formation during bone healing.

6.5.5 Platelet-Rich Plasma and Platelet-Rich Fibrin

Platelet-rich plasma (PRP) is a plasma fraction of autologous blood containing platelet concentration above the blood baseline level (at least 250,000 platelets per microliter). PRP naturally contain various endogenous growth factors, including PDGF, VEGF, and TGF-b. They improve tissue healing and regeneration by promoting new capillary growth, and improving host defense through immediate recruitment of neutrophils and macrophages, fibroblasts, and endothelial cells.

Platelet-rich fibrin (PRF) is a second-generation autologous platelet preparation and can be used in conjunction with MSCs and other alloplastic tissue constructs for simultaneous hard and soft tissue engineering in the maxillofacial region (e.g., alveolar bone, mandible, calvarium) (Li et al., 2014).

6.5.6 Transforming Growth Factor-β

Transforming growth factor-β3 plays a great role in developing palate by affecting palatal epithelial/mesenchymal transformation. Also, its supplementation has been used in surgical repair to promote scarless cleft lip repair.

6.6 CONCLUSION

This chapter outlines the various materials available and steps involved in replacement/regenerative therapy of craniofacial deformities using bone tissue engineering. Craniofacial deformities are debilitating as they affect the psychosocial, emotional, and functional well-being of the individual. These defects are difficult to treat due to the complexity of the craniofacial structures and their functional demands. However, with recent advancements in technology, bone tissue engineering serves as a promising alternative to the conventional methods of bone reconstruction/regeneration. Rapid advances in the development of biodegradable scaffolds and 3D printing technologies have expanded the realm of treatment choices available to these patients. With the advent of 3D technologies, automated imaging techniques can be used for imaging the defect followed by designing of a customized scaffold matrix using computer-aided designing software and incorporating the scaffold matrix with cells and biologically active signaling molecules to enhance the regenerative ability of the tissue constructs. Perhaps, to date, this area still remains relatively new and requires further in vivo and clinical trials for assessing its feasibility and effectivity, which has already been done in vitro.

REFERENCES

Aichelmann-Reidy, M. E., & Reynolds, M. A. (2008). Predictability of clinical outcomes following regenerative therapy in intrabony defects. *Journal of Periodontology*, *79*(3). https://doi.org/10.1902/jop.2008.060521

Al-Ahmady, H. H., Abd Elazeem, A. F., Bellah Ahmed, N. E., Shawkat, W. M., Elmasry, M., Abdelrahman, M. A., & Abderazik, M. A. (2018). Combining autologous bone marrow mononuclear cells seeded on collagen sponge with Nano Hydroxyapatite, and platelet-rich fibrin: Reporting a novel strategy for alveolar cleft bone regeneration. *Journal of Cranio-Maxillofacial Surgery*, *46*(9). https://doi.org/10.1016/j.jcms.2018.05.049

Assad, M., Chernyshov, A., Leroux, M. A., & Rivard, C. H. (2002). A new porous titanium-nickel alloy: Part 1. Cytotoxicity and genotoxicity evaluation. *Bio-Medical Materials and Engineering*, *12*(3), 225–237.

Bhumiratana, S., & Vunjak-Novakovic, G. (2012). Concise review: Personalized human bone grafts for reconstructing head and face. *Stem Cells Translational Medicine*, *1*(1). https://doi.org/10.5966/sctm.2011-0020

Carreira, A. C., Lojudice, F. H., Halcsik, E., Navarro, R. D., Sogayar, M. C., & Granjeiro, J. M. (2014). Bone morphogenetic proteins: Facts, challenges, and future perspectives. *Journal of Dental Research*, *93*(4). https://doi.org/10.1177/0022034513518561

Dorozhkin, S. V., & Epple, M. (2002). Biological and medical significance of calcium phosphates. *Angewandte Chemie- International Edition*, *41*(17). https://doi.org/10.1002/1521-3773(20020902)41:17<3130::AID-ANIE3130>3.0.CO;2-1

Doshi, J., & Reneker, D. H. (1995). Electrospinning process and applications of electrospun fibers. *Journal of Electrostatics*, *35*(2–3). https://doi.org/10.1016/0304-3886(95)00041-8

Ducheyne, P., & Qiu, Q. (1999). Bioactive ceramics: The effect of surface reactivity on bone formation and bone cell function. *Biomaterials*, *20*(23–24). https://doi.org/10.1016/S0142-9612(99)00181-7

Elsten, E. E. C. M., Caron, C. J. J. M., Dunaway, D. J., Padwa, B. L., Forrest, C., & Koudstaal, M. J. (2020). Dental anomalies in craniofacial microsomia: A systematic review. *Orthodontics and Craniofacial Research*, *23*(1). https://doi.org/10.1111/ocr.12351

Farré-Guasch, E., Wolff, J., Helder, M. N., Schulten, E. A. J. M., Forouzanfar, T., & Klein-Nulend, J. (2015). Application of additive manufacturing in oral and maxillofacial surgery. *Journal of Oral and Maxillofacial Surgery*, *73*(12). https://doi.org/10.1016/j.joms.2015.04.019

Fielding, G. A., Bandyopadhyay, A., & Bose, S. (2012). Effects of silica and zinc oxide doping on mechanical and biological properties of 3D printed tricalcium phosphate tissue engineering scaffolds. *Dental Materials*, *28*(2). https://doi.org/10.1016/j.dental.2011.09.010

Fuchs, J. R., Nasseri, B. A., & Vacanti, J. P. (2001). Tissue engineering: A 21st century solution to surgical reconstruction. *Annals of Thoracic Surgery*, *72*(2). https://doi.org/10.1016/S0003-4975(01)02820-X

Ghassemi, T., Shahroodi, A., Ebrahimzadeh, M. H., Mousavian, A., Movaffagh, J., & Moradi, A. (2018). Current concepts in scaffolding for bone tissue engineering. *Archives of Bone and Joint Surgery*, *6*(2). https://doi.org/10.22038/abjs.2018.26340.1713

Gładysz, D., & Hozyasz, K. K. (2015). Stem cell regenerative therapy in alveolar cleft reconstruction. *Archives of Oral Biology*, *60*(10). https://doi.org/10.1016/j.archoralbio.2015.07.003

Hermawan, H., Alamdari, H., Mantovani, D., & Dubé, D. (2008). Iron-manganese: New class of metallic degradable biomaterials prepared by powder metallurgy. *Powder Metallurgy*, *51*(1). https://doi.org/10.1179/174329008X284868

Ishaug, S. L., Yaszemski, M. J., Bizios, R., & Mikos, A. G. (1994). Osteoblast function on synthetic biodegradable polymers. *Journal of Biomedical Materials Research*, *28*(12). https://doi.org/10.1002/jbm.820281210

Lam, A. T. L., Reuveny, S., & Oh, S. K. W. (2020). Human mesenchymal stem cell therapy for cartilage repair: Review on isolation, expansion, and constructs. *Stem Cell Research*, *44*. https://doi.org/10.1016/j.scr.2020.101738

Li, Q., Reed, D. A., Min, L., Gopinathan, G., Li, S., Dangaria, S. J., Li, L., Geng, Y., Galang, M. T., Gajendrareddy, P., Zhou, Y., Luan, X., & Diekwisch, T. G. H. (2014). Lyophilized Platelet-Rich Fibrin (PRF) promotes craniofacial bone regeneration through Runx2. *International Journal of Molecular Sciences*, *15*(5). https://doi.org/10.3390/ijms15058509

Martín-Del-Campo, M., Rosales-Ibañez, R., & Rojo, L. (2019). Biomaterials for cleft lip and palate regeneration. *International Journal of Molecular Sciences*, *20*(9). https://doi.org/10.3390/ijms20092176

Merceron, T. K., Burt, M., Seol, Y. J., Kang, H. W., Lee, S. J., Yoo, J. J., & Atala, A. (2015). A 3D bioprinted complex structure for engineering the muscle-tendon unit. *Biofabrication*, *7*(3). https://doi.org/10.1088/1758-5090/7/3/035003

Nakajima, K., Kunimatsu, R., Ando, K., Ando, T., Hayashi, Y., Kihara, T., Hiraki, T., Tsuka, Y., Abe, T., Kaku, M., Nikawa, H., Takata, T., Tanne, K., & Tanimoto, K. (2018). Comparison of the bone regeneration ability between stem cells from human exfoliated deciduous teeth, human dental pulp stem cells and human bone marrow mesenchymal stem cells. *Biochemical and Biophysical Research Communications*, *497*(3). https://doi.org/10.1016/j.bbrc.2018.02.156

Nyberg, E. L., Farris, A. L., Hung, B. P., Dias, M., Garcia, J. R., Dorafshar, A. H., & Grayson, W. L. (2017). 3D-printing technologies for craniofacial rehabilitation, reconstruction, and regeneration. *Annals of Biomedical Engineering*, *45*(1). https://doi.org/10.1007/s10439-016-1668-5

Oliver, J. D., Jia, S., Halpern, L. R., Graham, E. M., Turner, E. C., Colombo, J. S., Grainger, D. W., & D'Souza, R. N. (2021). Innovative molecular and cellular therapeutics in cleft palate tissue engineering. *Tissue Engineering - Part B: Reviews*, *27*(3). https://doi.org/10.1089/ten.teb.2020.0181

Pourebrahim, N., Hashemibeni, B., Shahnaseri, S., Torabinia, N., Mousavi, B., Adibi, S., Heidari, F., & Alavi, M. J. (2013). A comparison of tissue-engineered bone from adipose-derived stem cell with autogenous bone repair in maxillary alveolar cleft model in dogs. *International Journal of Oral and Maxillofacial Surgery*, *42*(5). https://doi.org/10.1016/j.ijom.2012.10.012

Ratner, B. D., Hoffman, A. S., Schoen, F. J., & Lemons, J. E. (2004). Biomaterials science: An introduction to materials in medicine. *San Diego, California*, 162–164.

Salgado, A. J., Coutinho, O. P., & Reis, R. L. (2004). Bone tissue engineering: State of the art and future trends. *Macromolecular Bioscience*, *4*(8). https://doi.org/10.1002/mabi.200400026

Staiger, M. P., Pietak, A. M., Huadmai, J., & Dias, G. (2006). Magnesium and its alloys as orthopedic biomaterials: A review. *Biomaterials*. https://doi.org/10.1016/j.biomaterials.2005.10.003

Tian, T., Zhang, T., Lin, Y., & Cai, X. (2018). Vascularization in craniofacial bone tissue engineering. *Journal of Dental Research*, *97*(9). https://doi.org/10.1177/0022034518767120

Wang, M. (2003). Developing bioactive composite materials for tissue replacement. *Biomaterials*, *24*(13). https://doi.org/10.1016/S0142-9612(03)00037-1

Witt, P., Cohen, D., Grames, L. M., & Marsh, J. (1999). Sphincter pharyngoplasty for the surgical management of speech dysfunction associated with velocardiofacial syndrome. *British Journal of Plastic Surgery*, *52*(8). https://doi.org/10.1054/bjps.1999.3168

Witte, F., Kaese, V., Haferkamp, H., Switzer, E., Meyer-Lindenberg, A., Wirth, C. J., & Windhagen, H. (2005). In vivo corrosion of four magnesium alloys and the associated bone response. *Biomaterials, 26*(17). https://doi.org/10.1016/j.biomaterials.2004.09.049

Zhang, Y., & Zhang, M. (2001). Synthesis and characterization of macroporous chitosan/calcium phosphate composite scaffolds for tissue engineering. *Journal of Biomedical Materials Research, 55*(3). https://doi.org/10.1002/1097-4636(20010605)55:3<304::AID-JBM1018>3.0.CO;2-J

Zhao, P., Gu, H., Mi, H., Rao, C., Fu, J., & Turng, L. sheng. (2018). Fabrication of scaffolds in tissue engineering: A review. *Frontiers of Mechanical Engineering, 13*(1). https://doi.org/10.1007/s11465-018-0496-8

Zisch, A. H., Lutolf, M. P., & Hubbell, J. A. (2003). Biopolymeric delivery matrices for angiogenic growth factors. *Cardiovascular Pathology, 12*(6). https://doi.org/10.1016/S1054-8807(03)00089-9

7 Cold Spray Additive Manufacturing
A New Trend in Metal Additive Manufacturing

Kannan Ganesa Balamurugan
IFET College of Engineering, Villupuram, India

G. Prabu
National Institute of Technology, Tiruchirappalli, India

CONTENTS

7.1   INTRODUCTION

Manufacturing is a vital component of engineering where it transforms the design to the real-time component. Manufacturing fabricates products with the support of manpower, machinery, materials, and tools. Basically, two types of manufacturing exist, one is processing operations and other is assembly operations. Processing operations have further branched to shaping, property enhancing, surface treatment, and surface modification processes. Shaping processes are responsible to transfer the design to the real-time products. The shaping processes can be further classified as subtractive and additive manufacturing (AM). In subtractive manufacturing, bulk materials in the standard geometries like rectangular blocks or circular rods are transformed as per the design. This transformation can be achieved through removing the excess materials from the bulk geometry.

Subtractive manufacturing is well with the advancement of the CNC technology. However, fabrication of products with complex geometries to near net shape is still challenging in subtractive manufacturing (Wong & Hernandez, 2012). Even in the near net shaping processes like casting and forging, a specific die setup is required. Fabrication of those dies expects the machining operation. Due to this limitation, the design freedom could not be achieved through subtractive manufacturing. This limitation can be overwhelmed with the AM methods. AM fabricates three-dimensional (3D) components by layer-by-layer deposition of materials (Ngo et al., 2018). The AM technologies adopt a wide range of materials like polymers, metals, and ceramics (Hegab, 2016). Existing AM technologies are fused deposition modelling (FDM), powder bed fusion, ink jet printing, stereolithography (SLA), direct energy deposition, and laminated object manufacture (LOM) (Stavropoulos & Foteinopoulos, 2018). The 3D printing or AM technology started with the SLA as rapid prototyping technology. SLA method photo polymerize the liquid resin by exposing to the ultraviolet light source (Salonitis, 2014). The FDM technique is a devoted polymer printing technology, especially for thermoplastics. The frequently used thermoplastics are poly lactic acid (PLA) and acrylonitrile-butadiene-styrene (ABS). The thermoplastic polymeric materials are in the form of filament (wire structure), which feed through a heating unit and the semi-molten state polymer is extruded from the nozzle and deposited on the machine table.

The final 3D component will be fabricated by layer by layer deposition (Mohamed et al., 2014). The powder bed fusion method involves selective sintering or melting of metal powders spread on the machine table. Each layer of powder bed selectively sintered or melted and fused to the next layer subsequently by the systematic decrement of the machine table (Utela et al., 2008). The direct energy deposition method (DED) uses a high-intensity laser or electron beam to melt the metal powders feed along the beam and the substrate simultaneously and fused them by rapid solidification (Gibson et al., 2015). The DED process is otherwise called laser-engineered net shaping (LENS™). Powder bed setup is not required in this DED process. The above-mentioned technologies are dominating the AM arena. Recently, the cold spray AM (CSAM) has been added as a feather to the crown of powder-based AM techniques. The CSAM technique exploits the cold spray method to deposit the metallic powders with a high velocity on the substrate layer by layer to build the components (Fan et al., 2020). This review has conducted a detailed review based on the process parameters, materials used, and the future scope of the CSAM.

7.2 CONSTRUCTION AND WORKING OF COLD SPRAY MACHINE

An in-home CSAM system majorly consists of powder feeder, control panel, heater, compressed air supply, de Laval type converging–diverging nozzle, and a rotating chuck to hold the substrate. Based on the gas pressure, this system is classified into high-pressure CSAM (>1 MPa) and low-pressure CSAM (<1 MPa). The schematic

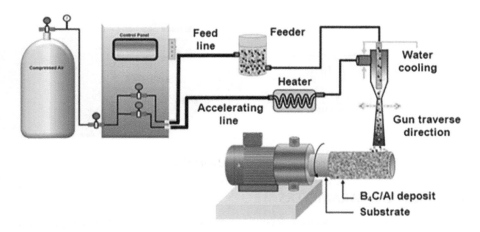

FIGURE 7.1 Schematic representation of high-pressure CSAM system.

representation of the high pressure CSAM system is shown in Figure 7.1 (Tariq et al., 2018). In this system, two gas lines are taken out form the compressor cylinder. One pipeline is passing through the feeder system, whereas other one is passing through the heater. The outlet of gas from the heater is named as accelerating gas or propulsive gas. Similarly, the gas from the feedstock system is called as carrier gas. The accelerating gas and carrier gas are thoroughly mixed, which leads to produce uniform mixing of the powder with accelerating gas. However, in this mixing process, the carrier gas pressure is set slightly higher than that of the accelerating gas. Thereby, the effectiveness of powders mixing is enhanced.

The resultant solid–gas mixture enters into the convergent–divergent de-Laval nozzle. The nozzle produced supersonic solid–gas phase stream due to its converging–diverging shape. The throat diameter, length of divergent section, and outlet diameter are 2 mm, 100 mm, and 6 mm, respectively (Luo et al., 2018). This medium-low-pressure supersonic steam is used to deposit the material on the substrates such as Al 6061 alloy (Schmidt et al., 2009). The supersonic velocity powder is deposited on the substrate by the robot-controlled spraying gun at the optimized spraying parameters. Supersonic velocity of the spraying medium is set such a way that the material deformed plastically and deposited on the substrate surface. Thereby, mechanical interlocking occurs between the deposited material with the substrate surface. It led to make a physical bond between these two materials (Assadi et al., 2003).

Moreover, high-pressure CSAM can also be used to deposit powder on metal surface at low pressure. The schematic representation of the low-pressure CSAM system is shown in Figure 7.2 (Yin & Lupoi, 2021). In the low-pressure CSAM system, powder is released on the substrate surface at the atmospheric pressure as the local surrounding pressure is maintained sufficiently lower than the atmospheric pressure. The low-pressure CSAM system is cheaper than the high-pressure system. However, processing parameters are limited in the low-pressure system than the high-pressure system. Hence, it is used to reconstruct the damaged components and deposit the low melting point material such as Cu and Al.

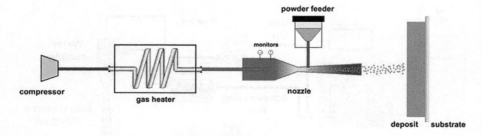

FIGURE 7.2   Schematic representation of low-pressure CSAM system.

7.3 PROCESS PARAMETERS

CSAM manufacturing processes depend on many parameters such as temperature, pressure, and type of aggressive gas, powder feed rate, scanning step, traverse speed, spray angle, standoff distance, and trajectory of nozzle. The properties of the deposit material depend on the particle velocity during the impact on the substrate surfaces. This particle velocity is controlled by the aggressive gas parameter. Hence, this gas parameter is considered as one of the important parameters in this CSAM manufacturing process. Different types of gases are used as the aggressive gas, namely, air, nitrogen, helium, etc. Impact velocity affects the efficiency of deposit material, mechanical characteristic, and bonding strength with the substrate. In addition to it, the hardness of the deposit coating is enhanced by increasing the velocity of the particle as it improves the work hardening effect.

Helium gas is used to increase the particle velocity and impact on the substrate surface. However, this high velocity induces the residual stress in the deposit material due to more plastic deformation occurring at high velocity. It can be overcome by depositing the material at higher temperature that reduces the residual stress by annealing effect. Air and nitrogen gases are used to reduce the cost of operating as the Helium gas has higher cost than other two gases. Moreover, the temperature and pressure of gas vary for different materials due to the variation in the melting point of the deposit material. The temperature and pressure of gas are in the range of 30 to 800°C and 1.0 to 5.0 MPa, respectively.

Powder particles' feed rate controls the mass of the particles supply into the nozzle section. More powder feed rate led to increase in the content of solid particles in the mixture and reduce the solid–gas mixture flow rate. It reduces the mechanical properties of the deposit material, increases the porosity, and induces more residual stress and delamination of material. Hence, the particle feed rate should be below 150 g/s. If the feed rate is very low, say 5 to 20 g/s, then the presence of the particle is less, which leads to reduction in the coating thickness. This powder feed rate along with the aggressive gas velocity affects the deposition profile of the material.

Nozzle travel speed controls the deposition of material per unit time and duration of the process. Lower travel speed allows more material to deposit on the target area and produce a thicker layer. If the nozzle travel speed varies, that influences the appearance of the surface and track thickness. Lower nozzle travel speed with higher powder feed rate produces non-uniform thickness of coating. Formation of track profile on the deposited region mainly depends on the flow velocity of accelerating gas. Significant differences occurred during the deposition of particles due to Gaussian distribution of the velocity profile. It affects the density, porosity, and mechanical properties such as adhesion strength and elastic modulus of the deposited material. In addition to it, pealing out of the deposited material occurred due to the low adhesive bonding strength. The low travel speed of nozzle heats the substrate and deposits the material to a higher temperature which leads to inducing the thermal stress between the substrate–deposit layer and layer–layer.

In the CSAM process, there is significant unfiled space between the two consecutive single-track layers. This space can be filled by using the scanning step parameter and make the resultant surface uniform. This width of the scanning step is equal to the distance between the centre lines of two neighbouring single-track deposits. The appropriate selection of the scan step leads to fabrication of the CSAM coating with uniform deposit thickness. In general, the scan step is taken as the half of the single-track width, which conformed the filling of the gap between the two neighbouring single tracks. The standoff distance of the nozzle is the distance between the target surface and the outlet edge of the nozzle which is another important parameter affecting the resultant CSAM surface. Solid–gas jet particles from the nozzle outlet are deviated from its target location because of resistance offered by the atmosphere.

Inside the nozzle, the solid powder particles have positive drag force and move along with the accelerating gas. However, this drags force of powder particles state to degrade and deviate from the accelerating gas velocity when it ejected from the nozzle outlet. As a result, the solid powder particles do not reach the target location at the substrate surface. Moreover, the impact force exerted by the powder particles is increased up to a certain value with the increase of the standoff distance. However, the impact force starts to reduce after the optimal distance. Hence, the standoff distance is selected in such a way that the impact force is maximum and deposits the powder on the target surface.

Deposition of powder particles on the target surface possessed two different components, namely, normal and tangential. Normal component supports the powder particles deposit on the target surface, whereas tangential component increases the sliding motion of the powder particles. As a result, the spray angle of the nozzle is optimized to increase the normal component and decrease the tangential component. These results lead to an increase in the adhesive bonding strength of the coating, deposit strength, and uniform profile cross-section of the deposit material.

7.4 PROPERTIES OF CSAM-FABRICATED SPECIMENS

The properties of the CSAM specimen mainly depend on the microstructure of the deposit layer. Hence, it is important to study the microstructure with the properties of the deposit for better understanding the involvement of the mechanism.

7.4.1 MICROHARDNESS

Microhardness of the CSAM coating is used to assess the withstand capacity of coating during the plastic deformation, scratching, and indentation by hard particles. Therefore, it is important for engineering standpoint as wear of the component is

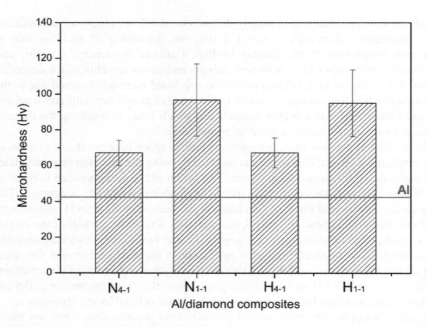

FIGURE 7.3 Microhardness profile of the CSAM Al/diamond composites, and pure Al.

inversely proportional to the hardness of the component. The microhardness properties of Al/diamond coating produced by the CSAM method is analysed and the corresponding data are plotted in Figure 7.3 (Chena et al., 2020). The results revealed that the microhardness of the Al/diamond CSAM specimens is higher than that of the Al CSAM specimens due to dispersive strengthening mechanism of the diamond element.

7.4.2 Tribological Property

CSAM coatings do not easily peel out during the wear test due to more adhesive strength of the coating. The tribological properties of Al/diamond CSAM are analysed, and it has been identified that the diamond particles are not worn out form the composites coating as shown in Figure 7.4 (Chena et al., 2020). This prevents the removal of the Al element and increases the durability of the coating for a longer time. The tribofilm contains W and O elements that are crushed during the sliding process. It resulted in the formation of the thick tungsten carbide layer on the wornout surface. It behaves as the protecting layer and prevents the Al/diamond underneath the content of the coated region. This WC protective layer thickness varied with respect to the accelerating gas type such as helium and nitrogen. As a result, the surface morphology of four wornout surface is significantly varied as shown in Figure 7.4 (Chena et al., 2020). Moreover, the contact area of the CSAM specimen with the mating part is varied with respect to the intensity of mixing elements as the N1–1 and H1–1 samples possessed lower wornout surface because of higher diamond contact, whereas N4–1and H4–1 samples experienced higher cracks because of low diamond contact.

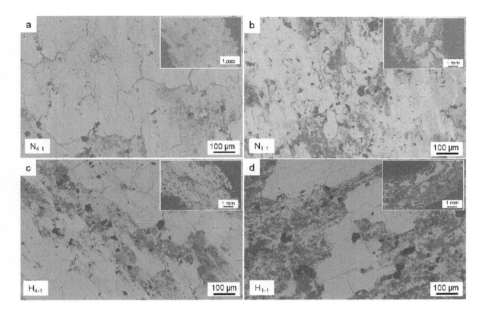

FIGURE 7.4   FESEM images of the wornout surfaces on the CSAM Al/diamond.

7.4.3   SURFACE ROUGHNESS

The surface roughness of the deposit layer affects the surface properties of the CSAM component during contact with the matting surface. It is found that the accelerating gas temperature affects the surface roughness of the deposit during the fabrication of Al CSAM coating (Jenkins et al., 2019).

There is no exact relationship for the accelerating gas temperature with the surface roughness. The surface roughness of Al coating initially decreased with the increase in the gas temperature up to 450°C and then started to increase with the increase in temperature as shown in Figure 7.5 (Jenkins et al., 2019). It is identified that the presence of pores on the surface of the Al CSAM-coated region might be the reason for increasing the surface roughness of coating.

7.4.4   THERMO-MECHANICAL PROPERTY

After the CSAM process, influence of the thermo-mechanical treatment on the microstructural formation is studied on the deposition of B_4C/Al elements on the Al6061 substrate, which is shown in Figure 7.6 (a–d) (Tariq et al., 2018).

The formation of substructure grains and distribution of recrystallized grains and deformed grains studied at different thermo-mechanical treatment are analysed and it was found that the substructure grains and distribution of recrystallized grains increased after the thermo-mechanical treatment. However, the deformed grains of the thermo-mechanical treatment samples (Figure 7.6b–d) are decreased when compared with the CSAM sample (Figure 7.6a). The numerical data of substructure grains, distribution of recrystallized grains, and deformed grains are plotted in a single graph as shown in Figure 7.6e.

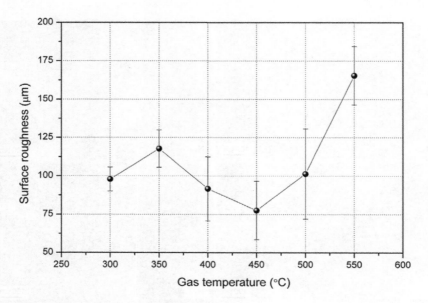

FIGURE 7.5   Surface roughness of the CSAM sample.

7.4.5 TENSILE STRENGTH

The stress–strain curves of the B_4C/Al CSAM composite coated as-sprayed and heat-treated at 200 to 500°C at an interval of 100°C are shown in Figure 7.7 (Tariq et al., 2018). The tensile strength of the CSAM specimen depends on the porosity, inter-splat bonding, work hardening, and grain size. A large number of porosities (3.9%) are present in the as-sprayed specimen that leads to premature failure of the component. The as-sprayed specimen fractured at the elastic region has 38 MPa tensile strength, which is the lowest strength among the as-prayed and heat-treated specimen. Moreover, the presence of weakly bonded inter-layer and inter-splat boundaries in the as-sprayed coating might reduce its tensile strengths. However, the 200°C heat-treated specimen possesses 44 MPa tensile strength, which shows slight improvement in the strength. This improvement in tensile strength is due to healing of the inter-splat boundaries and removal of localized stresses. In addition to it, the increase in heat treatment temperature refines the grain size and heals the inter-splat boundaries. As a result, the ductility of the specimen at 300°C, 400°C, and 500°C is higher than that of the as-sprayed specimen. Corresponding ductility values are increased to 0.7, 1.1, and 1.4%, respectively. The yield strength of the heat-treated specimens decreases but the ultimate tensile strength retains the same values relatively. Moreover, the higher temperature reduces the inter-splat boundary defects by the diffusion of the inherent elements that reduce the degradation properties. As a result, the tensile strength and ductility of the heat-treated specimen are further improved.

The tensile property of the CSAM specimen can be enhanced by post-processing treatment. The presence of pores on the CSAM specimen may be reduced by hot isostatic (HIP) post-process treatment. The low operating temperatures of the CSAM

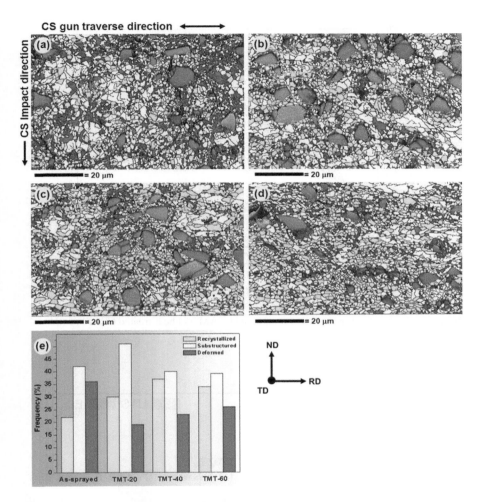

FIGURE 7.6 EBSD map (a) as-sprayed CSAM specimen, (b) TMT-20 specimen, (c) TMT-40 specimen, (d) TMT-60 specimen, and (e) Panel – grains at various processing conditions.

process result in degrading the mechanical properties of the coated material when compared with its fusion AM for high-strength 316L stainless steel materials. The tensile strength of the 316L steel CSAM specimens before and after the annealing treatments is shown in Figure 7.8 (Yin et al., 2019).

The tensile strength of CSAM specimens before the HIP process is less than the after-HIP process because of the presence of pores and weak inter-layer bonding. After the HIP process, the elongation and tensile strength increase due to recrystallization and improvement in bonding between inter-particle layers as shown in Figure 7.9. During the HIP process, the oxide thin layers are compactedly deposited on pores. The pores are collapsed and form metallurgical bonding between layers as shown in Figure 7.9 (Shuo Yin et al., 2019).

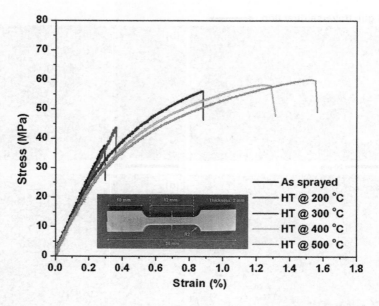

FIGURE 7.7   Stress–strain curve of as-sprayed CSAM and heat-treated CSAM samples.

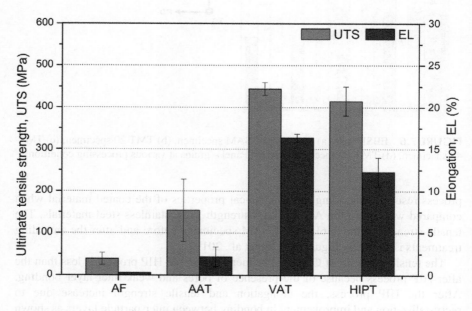

FIGURE 7.8   UTS and elongation of the 316L CSAM deposits and after annealing treatments.

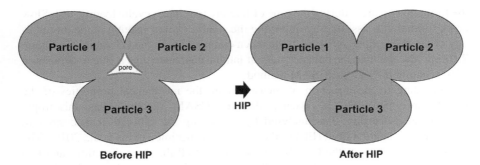

FIGURE 7.9 Schematic representation of inter-particle pore.

7.4.6 FRACTURE TOUGHNESS

The fracture toughness of heat-treated (HT) and double heat-treated (DT) cold-sprayed Cr_3C_2-Ni coating is shown in Figure 7.10 (Jafarlou et al., 2018). Cracks are slowly grown in the CSAM specimen due to shear lips and catastrophic growth in the rough-faced region (Figure 7.10a and b). The presence of shear slips in the fracture surface confirmed the transfer of failure mechanism from brittle facture to ductile facture. Compared with the double heat treatment sample, the single heat treatment sample has 7 PMa higher fracture toughness because of grain refinement. Porosity and defects are also the reason for the presence of discontinuities that alter

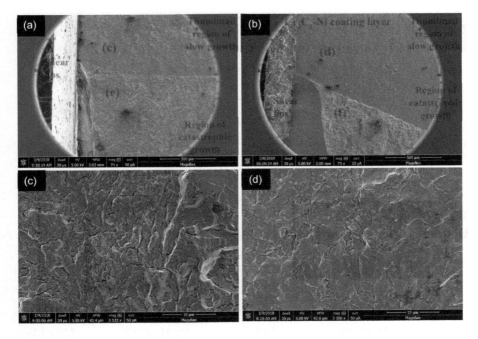

FIGURE 7.10 Fracture toughness test (a) HT, (b) DT, (c) magnified view of HT, and (d) magnified view of DT.

the formation of stress during the time of fracture toughness cyclic loading test. Slow crack formation is assessed by observing the small striation spacing in the morphology of the HT sample in Figure 7.10c when compared with the DT sample in Figure 7.10d. It revealed that the enhancement of fatigue characteristics is due to deposition of the coating through the CSAM method.

According to the direction of measurement, the mechanical properties of the CSAM specimen vary. It revealed that the CSAM specimen has anisotropic mechanical properties. It is analysed by measuring the ultimate tensile strength (UTS) and elongation limit (EL) of the Cu CSAM specimen (Yina et al., 2018). The UTS and EL values of the Cu CSAM specimen at all the three directions and two different built directions are listed in Tables 7.1 and 7.2. The result revealed that the Cu CSAM specimens exhibit anisotropy property at both the built directions. The UTS value of the bidirectional strategy Cu CSAM specimen at X direction has higher value than other two directions as well as cross hatch strategy specimen. In contrast, the UTS value of the cross hatch strategy Cu CSAM specimen at X direction has a much lower value than the other two directions.

These values confirmed that the Cu CSAM specimen possesses anisotropy mechanical property. Moreover, the mechanical property of the Cu CSAM specimen is significantly reduced after annealing process due to modification in the grain structure. This variation in the mechanical properties suggests that the cross-hatching strategy specimen is preferred for obtaining the isotropic properties throughout the specimen compared with bidirectional strategy. The reason for the variation is

TABLE 7.1
UTS, EL of CSAM Specimen at Bidirectional Strategy (Yin et al., 2018)

	Bidirectional Strategy			
	As Fabricated		Annealed	
	UTS (MPa)	EL (%)	UTS (MPa)	EL (%)
X	178.10 ± 14.41	2.43 ± 0.32	219.48 ± 11.55	8.86 ± 0.97
Y	109.71 ± 22.53	1.75 ± 0.44	144.71 ± 10.88	6.73 ± 1.01
Z	133.12 ± 5.92	2.89 ± 0.15	156.78 ± 11.11	7.6 ± 1.00

TABLE 7.2
UTS, EL of CSAM Specimen at Cross-Hatching Strategy (Yin et al., 2018)

	Cross-hatching Strategy			
	As Fabricated		Annealed	
	UTS (MPa)	EL (%)	UTS (MPa)	EL (%)
X	116.54 ± 8.53	2.08 ± 0.35	188.81 ± 16.31	8.37 ± 0.87
Y	120.69 ± 15.60	2.72 ± 0.59	168.94 ± 21.92	7.29 ± 1.29
Z	139.42 ± 5.99	2.98 ± 0.30	159.63 ± 7.13	6.95 ± 1.00

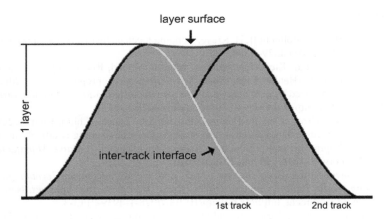

FIGURE 7.11   Schematic representation of a single-track deposit.

clarified in Figure 7.11 (Yin et al., 2019), which contains two single track Cu deposit region in the XY plane. The deposit shape of the first track is in the form of pyramid and the second track is deposit adjustment to the first track with some overlap ration to obtain the uniform shape. As a result, the second track is deposited at an angle with the first track. It leads to a change in the property of the specimen with respect to the direction of the deposition. Moreover, standing step deposition of the material is performed between the two tracks for obtaining a uniform flat structure. Hence, an intertrack layer is formed between the two tracks that had less bond strength compared with adjustment of both tracks. These intertrack layers fail during the tensile test, which led to grow the anisotropy property of material. In addition, accumulation of such steeping set inter-layer results in lowering the UTS and EL properties of the material.

7.5   CONCLUSION

In this chapter, the process parameter and mechanical property of CSAM specimens are analysed. However, the correlation between the CSAM mechanical property and the microstructure is not explored properly. Moreover, the mechanism involved in the formation of different microstructure and deformation with respect to different mechanical tests are not focussed thoroughly. These analyses should include all types of material such as ceramics, metals, polymers, and other nanosize powders. In addition, the investigation involving quantitative relation to the process parameters and characteristic of the materials is not available in open literature.

   In future work, studies will be mainly focussed on investigating the microstructure of the CSAM coating and correlating to basic mechanical properties that will improve the understanding of the concept involved in this process. In addition, the response of the CSAM specimen to practical applications such as wear, fatigue/fretting wear, corrosion, and erosion is an important requirement to be studied in order to apply the present CSAM technology to real application. The future efforts may help the CSAM technology to use in new applications and will reduce the cost of manufacturing.

REFERENCES

Assadi H., F. Gärtner, T. Stoltenhoff, H. Kreye, Bonding mechanism in cold gas spraying, *Acta Materialia* 51 (2003) 4379–4394.

Chen, Chaoyue, Yingchun Xie, Xingchen Yan, Mansur Ahmed, Rocco Lupoi, Jiang Wang, Zhongming Ren, Hanlin Liao, Shuo Yin, Tribological properties of Al/diamond composites produced by cold spray additive manufacturing, *Additive Manufacturing* 36 (2020) 101434, https://doi.org/10.1016/j.addma.2020.101434

Fan, Ningsong, Jan Cizek, Chunjie Huang, Xinliang Xie, Zdenek Chlup, Richard Jenkins, Rocco Lupoi, Shuo Yin, (2020) A new strategy for strengthening additively manufactured cold spray deposits through in-process densification, *Additive Manufacturing*, https://doi.org/10.1016/j.addma.2020.101626

Gibson, Ian, David Rosen, Brent Stucker (2015) *Additive Manufacturing Technologies, 3D Printing, Rapid Prototyping, and Direct Digital Manufacturing*, Springer, New York, Heidelberg Dordrecht, http://dx.doi.org/10.1007/978-1-4939-2113-3

Hegab, Hussien A., (2016) Design for additive manufacturing of composite materials and potential alloys: A review, *Manufacturing Review* 3 11, https://doi.org/10.1051/mfreview/2016010

Jafarlou, Davoud M., Caitlin Walde, Victor K. Champagne, Sundar Krishnamurty Ian R. Grosse, (2018) Influence of cold sprayed Cr3C2-Ni coating on fracture characteristics of additively manufactured 15Cr-5Ni stainless steel, *Materials and Design* 155 (2018) 134–147, https://doi.org/10.1016/j.matdes.2018.05.063

Jenkins, Richard, Barry Aldwell, Shuo Yin, Subhash Chandr, Gary Morgan, Rocco Lupoi, (2019) Solid state additive manufacture of highly-reflective Al coatings using cold spray, *Optics and Laser Technology* 115 251–256, https://doi.org/10.1016/j.optlastec.2019.02.011

Luo, Xiao-Tao, Meng-Lin Yao, Ninshu Ma, Makoto Takahashi, Chang-Jiu Li (2018) Deposition behavior, microstructure and mechanical properties of an in-situ micro-forging assisted cold spray enabled additively manufactured Inconel 718 alloy, *Materials and Design* 155 (2018) 384–395, https://doi.org/10.1016/j.matdes.2018.06.024

Mohamed, Omar A., Syed H. Masood, Jahar L. Bhowmik, (2014) Optimization of fused deposition modeling process parameters: A review of current research and future prospects, *Advances in Manufacturing* http://dx.doi.org/10.1007/s40436-014-0097-7

Ngo, Tuan D., Alireza Kashani, Gabriele Imbalzano, Kate T. Q. Nguyen, David Hui (2018) Additive manufacturing (3D printing): A review of materials, methods, applications and challenges, 208760, https://doi.org/10.1016/j.compositesb.2018.02.012

Salonitis K., (2014) Stereolithography, *Comprehensive Materials Processing*, 10 http://dx.doi.org/10.1016/B978-0-08-096532-1.01001-3

Schmidt T., H. Assadi, F. Gärtner, H. Richter, T. Stoltenhoff, H. Kreye, T. Klassen, (2009) From particle acceleration to impact and bonding in cold spraying, *Journal of Thermal Spray Technology* 18 794–808.

Stavropoulos, Panagiotis, Panagis Foteinopoulos, (2018) Modelling of additive manufacturing processes: A review and classification, *Manufacturing Review* 5 (2), https://doi.org/10.1051/mfreview/2017014

Tariq, N. H., L. Gyansah, X. Qiu, H. Dua, J. Q. Wang, B. Feng, D. S. Yan, T. Y. Xiong, (2018) Thermo-mechanical post-treatment: A strategic approach to improve microstructure and mechanical properties of cold spray additively manufactured composites, *Materials and Design* 156 (2018) 287–299, https://doi.org/10.1016/j.matdes.2018.06.062

Utela, Ben, Duane Storti, Rhonda Anderson, Mark Ganter (2008) A review of process development steps for new material systems in three dimensional printing (3DP), *Journal of Manufacturing Processes* 10, 96–104, http://dx.doi.org/10.1016/j.jmapro.2009.03.002

Wong, Kaufui V., Aldo Hernandez (2012). A Review of additive manufacturing. *International Scholarly Research Network*, 208760, https://doi.org/10.5402/2012/208760

Yin, Shuo, Rocco Lupoi, (2021) Cold spray additive manufacturing, *Springer Tracts in Additive Manufacturing*, https://doi.org/10.1007/978-3-030-73367-4

Yina, Shuo, Jan Cizek, Xingchen Yan, Rocco Lupoi (2019) Annealing strategies for enhancing mechanical properties of additively manufactured 316L stainless steel deposited by cold spray, *Surface & Coatings Technology* 370 (2019) 353–361 https://doi.org/10.1016/j.surfcoat.2019.04.012

Yina, Shuo, Richard Jenkins, Xingchen Yan, Rocco Lupoi, (2018) Microstructure and mechanical anisotropy of additively manufactured cold spray copper deposits, *Materials Science & Engineering A* 734 (2018) 67–76, https://doi.org/10.1016/j.msea.2018.07.096

8 Development and Mechanical Characterization of Coir Fiber-Based Thermoplastic Polyurethane Composite

Jaspreet Kaur and Dharmpal Deepak
Punjabi University, Patiala, India

Harnam Singh Farwaha
Guru Nanak Dev Engineering College, Ludhiana, India

Sulakshna Dwivedi
University School of Applied Management, Punjabi
University, Patiala, India

Nishant Ranjan
University Centre for Research and Development,
Chandigarh University, Mohali, India

CONTENTS

DOI: 10.1201/9781003266464-8

8.1   INTRODUCTION

Thermoplastic polyurethane (TPU) is one of the course groups of polyurethane plastics with various properties, such as transparency, elasticity and resistance oil and abrasion, and grease (Behniafar and Azadeh, 2015). Scientifically, thermoplastic elastomers contain linear segmented block copolymers poised of soft and hard segments (Yuan et al., 2019). This is partly because TPU is a linear segment block polymer composed of smooth and complex components. Its hard element can either be aromatic or aliphatic (Mi et al., 2017). Aromatic TPUs are based on isocyanates like methylene diphenyl diisocyanate (MDI), whereas aliphatic TPUs are based on isocyanates like H 12 MDI. When these isocyanates are combined with short-chain diols, they become the hard block. Generally, it is aromatic, but retention in sunlight exposure is a priority by color and clarity, an aliphatic hard segment is often used (Gryshchuk, 2005).

Depending on the application, the soft segment can either be a polyether or polyester type. For example, wet environments require a polyether-based TPU, whereas oil and hydrocarbon resistance often demand a polyester-based TPU (Chattopadhyay and Raju, 2007; Magnin et al., 2020). Table 8.1 shows the comparison of the physical properties of different natural fibers.

TABLE 8.1
Comparison of Physical Properties of Abaca with Other Natural Fibers

Physical Properties	Abaca	Hemp	Jute	Sisal	Linen	Cotton	Coir
Density (g/cm³)	1.5	1.48	1.46	1.33	1.4	1.54	1.2
Fiber length	2–4 m	1–2 m	3–3.5 m	1 m	Up to 90 cm	10–65 mm	200 μm
Fiber diameter	150–260 Microns	16–50 Microns	60–110 Microns	100–300 microns	12–60 microns	11–22 Microns	—
Tensile strength (N/m²)	980	550–900	400–800	600–700	800	400	144.6 N/m²
Elongation	1.1%	1.6%	1.8%	4.3%	2.7–3.5%	3–10%	32.3%
Moisture regain	5.81%	12%	13.75%	11%	10–12%	8.5%	—
Young's modulus (GPa)	41	30–60	20–25	17–22	50–70	6–10	3101.2

8.1.1 CHEMICAL CLASSIFICATION OF TPU

8.1.1.1 POLYESTER TPUs

TPU is familiar for its excellent abrasion resistance. Polyester-based TPU is used in applications requiring excellent resistance to fuels, oil, water, and other chemicals (Hu et al., 2020). Polyester TPUs typically offer higher-level transparency.

8.1.1.2 Polyether TPUs

TPUs are familiar for their excellent abrasion resistance. Polyether-based TPUs are used where resistance to moisture, water, and microorganisms is needed (Moerman and Partington, 2014). The specific gravity of polyether TPUs is marginally lower than that of polyester and polycaprolactone grades. They offer superior abrasion and tear resilience and present low-temperature flexibility (Harynska et al., 2019). Polyether TPUs are acceptable for those applications where moisture is in consideration.

8.1.1.3 Polycaprolactone TPUs

They are the consummate basic material for pneumatic and hydraulic seals. Polycaprolactone TPUs display inherent toughness and relatively high resistance to hydrolysis (Johnson and Samms, 1997).

8.1.2 CLASSIFICATION OF TPUs BASED ON AROMATIC AND ALIPHATIC VARIETIES

8.1.2.1 AROMATIC TPUs

They are based on isocyanates like MDI, which are workhorse products and can be used in applications where flexibility, toughness, and strength are required. The aromatic coating can break down over time if left in the sun. For any products that spend most of their usable lives outdoors, which include inflatable boats and flexible storage tanks, it is extensively used.

8.1.2.2 Aliphatic TPUs

Aliphatic TPUs do not oxidize under UV radiation, making it one of the much common coating compounds available for outdoor applications like color stability upon UV exposure and non-yellowing appearance. Aliphatic TPU has the ability to withstand oxidation from sunlight.

8.1.3 PHYSICAL PROPERTIES OF THERMOPLASTIC POLYURETHANE

8.1.3.1 SHORE HARDNESS

Thermoplastic elastomers are calculated inshore A and Shore D hardness according to ISO868. Shore hardness is computed from resistance of a material to the penetration of needle under distinct spring force. It is firm as a number on or after 0 to 100 on the scale A or D. The higher the number, the superior the hardness. Letter A is used for flexible types and letter D for rigid types. However, the ranges do partly cover. This chart illustrates a contrast of the shore hardness A and D scales for A pilon 52 TPU materials. Shore hardness decreases as temperature increases. Figure 8.1 shows the correlation graph of Shore A vs Shore D hardness.

8.1.3.2 Tensile Strength

It indicates TPU behavior when a specimen is placed under the short term, uniaxial stress.

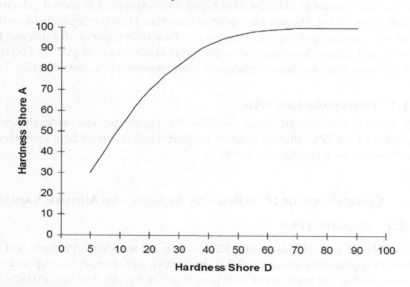

FIGURE 8.1 Correlation between Shore A and D.

8.1.3.3 Tear Strength

Tear strength is the term that defines the resistance of a notched specimen (TPU elastomers) to tear propagation. Tear strength is generally regarded as the critical indicator of the physical strength of the external force. In this respect, our film is superior to most other plastics.

8.1.3.4 Compression Set

Compression set is often a property of interest when using elastomers. A compression set is the amount of permanent deformation that occurs when a material is compressed in a specific deformation, for a specified time, at a specific temperature. ASTM D395 Standard Test Method for Rubber Property – compression set is the test method used and it calls for the material to be 25% deformed for a given period. After a 30-min recovery time, the sample measurement is undertaken. The value derived is the proportion a material sample fails to recover from its original height.

8.1.3.5 Abrasion

Coarse paper is applied to a substrate under pressure via a rotating cylinder to measure the abrasion resistance of plastic materials like TPU. Before and after the abrasion assessment, the weight of the specimen is measured. The original density of the material is considered alongside the roughness of the paper with results typically expressed in terms of volume loss of the substrate in mm^3.

8.1.3.6 Shrinkage

Different factors influenced the shrinkage of TPU such as wall thickness, part design, processing condition, gate design, melt and mold temperature as well as injection and holding pressure. Figure 8.2 shows the shrinkage ratio (in %age) for TPU grades according to wall thickness and hardness.

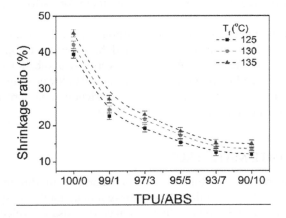

FIGURE 8.2 Total shrinkage for TPU grades in relation to wall thickness and hardness.

8.2 NATURAL FIBER

Natural fiber is considered recyclable and may have potential for some applications. These days, this type of fiber has to gain popularity as a filler in polymer composites. Apart from this, it is readily available and environmentally friendly. It is also inexpensive because they are non-toxic and biodegradable waste (Li et al., 2020). This chapter presents various natural fibers from fruit peel, fruit skin, waste, agricultural residue, and others as fillers in TPU (Ahad et al., 2020). The author also reflects the key issue related to the effect of filler loading on the tensile properties of TPU composites. In general, the tensile strength of the natural fiber-filled polymer composites increases with fiber loading. The tensile properties are largely influenced by the interfacial bond between the polymer and the fibers. Other than that, each type of fiber also plays a vital role in its properties and polymer adhesion. This re-examination barely focuses on the most recent studies in the array of 7 years commencing from 2013 to 2019 (Saeed et al., 2021).

These days, natural fiber is an interesting option and it is the most widely applied fiber in composite technology. Natural fibers are used in most industries and are increasingly being considered as reinforcements for polymer matrix composites because they are perceived to have sustainability value (Norhidayah et al. 2014).

Recently, there has been a rapid growth in research and innovation in the natural fiber composite area. Interest is warranted due to the advantages of these materials compared to others, such as synthetic fiber composites, including low environmental impact and low cost and support their potential across a wide range of applications. Much effort has gone into increasing their mechanical performance to extend the capabilities and applications of this group of materials (Pickering et al., 2016).

Fiber-reinforced composites were in use since ancient times. Due to the drawback of synthetic and fiberglass as reinforcement, the use of fiber-reinforced composite increased the interest of the researcher or the scientist. This review article investigates the use of various fibers as reinforcement in composites. With the innovation in science and technology, the new means of depiction and evaluation of physical, chemical, thermal, and mechanical properties of the composite have been used that have explored the new scope of using them for various purposes (Ranga et al., 2014).

The present experimental study aims at learning the mechanical behavior of natural fiber composites. Natural fibers have attracted the attention of engineers, researchers, professionals, and scientists all over the world as a substitute reinforcement because of their outstanding properties such as low cost, lower weight, high specific strength, mechanical properties, fairly good, eco-friendly, non-abrasive, and biodegradable characteristics (Elanchezhian et al., 2018).

8.3 COMPOSITE MATERIALS

A composite material is a combination of two materials with different physical and chemical properties. When combined, they make a material that is specific to a certain work, for instance, to become lighter, stronger, or resistant to electricity and also with better strength and stiffness.

8.3.1 COMPOSITE MATERIALS PHASES

(a) **Primary phase**

The primary phase forms the matrix within which the second phase is embedded. Secondary phase – embedded phase sometimes referred to as a reinforcing agent, because it usually serves to strengthen the composite. The reinforcing phase may be in the form of fibers, particles, or various other geometries.

(b) **Secondary phase**

This phase is to be embedded into the primary phase which tends to strengthen the composite. It may contain fibers and particles.

8.3.2 TYPES OF COMPOSITE MATERIALS

There are five types of composite materials:

(a) Fiber composite
(b) Particle composite
(c) Flake composite
(d) Laminar composite
(e) Filled composite

8.3.3 CLASSIFICATION OF COMPOSITE MATERIALS

(a) Metal matrix composites
(b) Ceramic matrix composites
(c) Polymer matrix composite

8.4 PROBLEM FORMULATION

From the last few decades, the applications of natural fiber-reinforced composites have increased from aerospace structure automotive parts and from building materials to sporting goods. Attention toward development and utilization of natural fiber has increased due to their high strength and modulus. These are increasingly cost effective, economical, and biodegradable as compared with synthesis fiber. The plant fiber provides excellent insulation against heat and noise.

The literature review reveals that composites with higher stiffness, light in weight, fatigue properties, and toughness could be achieved by the use of natural fibers; abaca fiber is found to be potential reinforcement in composites and it also possesses higher tensile strength and young's modulus. A lot of research has been done to discover the suitability of abaca fiber as reinforcement in polymer composites to produce more economical polymer composites.

Recycling of plastic has become the biggest challenge for engineers because of their extensive use in daily life. Some of the composites are produced now a days with biodegradable properties which would help the world to decompose the waste plastic.

After reviewing the literature, it has been found that limited work has been done using coir fiber as reinforced in the matrix of thermoplastic polyurethane. Most of the research is based on polyethylene and polypropylene matrices used in the fabrication of the composite. Hence, it has been decided to fabricate coir/TPU composite. The effects of varying the fiber content on the mechanical properties of coir/TPU composites have been investigated.

8.5 EXPERIMENTATIONS

In this study, coir fiber is used as reinforced and TPU is used as the matrix. Coir fiber was procured from Ecogreen Chennai, Tamil Nadu, and TPU of U92 Covestro grade from Rai Innotech Polymers Pvt. Ltd, New Delhi.

8.5.1 CHEMICAL TREATMENT OF COIR FIBER

Chemical treatment of the reinforcement material is one of the main ways of improving the mechanical properties of natural fiber-reinforced polymer composites. In the present study, coir fiber was used as reinforcement material, whereas polypropylene (PP) and polyethylene (PE) polymer were used as the matrix material. Before reinforcing with polymer, raw coir fiber was chemically treated with basic chromium sulphate and sodium bicarbonate in a sieve shaker. Hot-pressed method was used for composite manufacturing during which fiber loading was varied at 0, 5, 10, 15, and 20 wt%. Comparison of the properties of raw and chemically treated coir fiber-reinforced PP and PE was conducted. Mechanical characteristics of the composites were evaluated using tensile, flexural, impact, and hardness tests. Water absorption test was conducted to know water uptake characteristics. Microstructural analysis using a scanning electron microscope was performed to observe the adhesiveness between the matrix and the fiber. Thermogravimetric analysis was done to observe the physical and chemical changes in fiber and composites. The results showed that chemical treatment improved the physical, mechanical, and thermal properties of the manufactured composites (Mir et al. 2015). Figure 8.3 shows the treatment of coir fiber in different stages.

8.5.2 FABRICATION OF TENSILE AND FLEXURAL SPECIMENS

To investigate the influence of varying contents of coir fiber on textile and flexural properties of coir fiber-reinforced thermoplastic polyurethane composite, samples were prepared by taking 10%, 20%, and 30% coir fiber by wt. in TPU. For comparison purpose, the samples were also prepared containing fresh TPU. The description of the different samples is given in Table 8.2. Fabricated TPU tensile and flexural samples are shown in Figure 8.4.

Where $\rho_f = 1.50$ g/cc and $\rho_m = 1.23$ g/cc are the densities of the coir fiber (Biswas and Satapathy, 2010).

Tensile and flexural specimens were fabricated by putting corresponding mixtures of the samples given in Table 8.2 into injection molding machine (Model: NG80, Make: Neelgiri).

(a)   (b)

(c)   (d)

FIGURE 8.3 (a) Washing of fiber. (b) Heating of fiber in oven. (c) Soaking of coir fiber in 8% wt NaOH. (d) Soaking of abaca coir in 10% wt acetone + MA.

TABLE 8.2
Composition of Tensile and Flexural Specimens

Sample No.	Sample Composition of Coir Fiber	(wt%) TPU	Sample Density
1	0	100	1.23
2	10	90	1.257
3	20	80	1.284
4	30	70	1.311

8.5.3 INVESTIGATING THE FAILURE MODE OF TENSILE SPECIMEN

To investigate the morphology of fractured tensile specimens and prevailing mode of failure, the failed surface was observed under a scanning electron microscope (SEM) [Model: JSM-IT500, Make-JEOL]. The fractured surface was coated with gold in DII-29030SCTR smart coater, before SEM investigation. Gold plating was done to enhance the electric conductivity of the fiber-reinforced polyurethane composite. Gold-plated samples were mounted on the stubs with silver paste and placed

FIGURE 8.4   Pure TPU specimen for tensile and flexural strength testing.

inside SEM machine in which vacuum was created to obtain a clear image under high magnification. The SEM helped in revealing the extent of bonding between coir fibers and the TPU matrix.

8.6   RESULTS AND DISCUSSION

8.6.1   Tensile Strength at Different Weight Percent of Coir Fiber in the TPU Matrix

From each of the fabricated composites, three similar specimens were taken and subjected to tensile tests. To represent the final tensile strength of the sample, the average of these three similar specimens was reported. ASTM D 638 standard were followed to perform tensile tests. The results obtained from various specimens are reported in Table 8.3 along with mean standard deviation.

Table 8.3 represents the tensile strength of the various specimens tested in the present study. The maximum variations in the tensile strength of specimens in samples T0, T1, T2, and T3 from the mean tensile strength of the lots are observed as 15.88, 12.76, 9.41, and 11.67 respectively, whereas the minimum variation in tensile strength of specimens in samples T0, T1, T2, and T3 from the mean tensile strength of samples is observed as −0.39, −1.47, 0.05, and 0.13, respectively. The comparison of standard deviations in tensile strength lots reveals the variation in tensile strength of specimens corresponding to sample T1 is the highest, whereas it is observed to be the lowest for sample T2.

8.6.2   Flexural Strength of Different Specimens

Similar to tensile tests, three similar specimens were taken and subjected to flexural tests to represent the final flexural strength of the sample, and average results of these

TABLE 8.3
Tensile Strength of Different Specimens

Specimen Notation	Specimen no.	Tensile Strength (MPa)	Mean Tensile Strength (MPa)	Standard Deviation (MPa)
T0	1	19.57	15.88	2.63
	2	14.44		
	3	13.61		
T1	4	12.01	12.76	1.85
	5	15.23		
	6	11.06		
T2	7	8.72	9.41	0.48
	8	9.81		
	9	9.67		
T3	10	12.66	11.67	1.31
	11	9.81		
	12	12.54		

TABLE 8.4
Flexural Strength of Different Specimens

Specimen Notation	Specimen No.	Flexural Strength (MPa)	Mean Flexural Strength (MPa)	Standard Deviation (MPa)
F0	1	6.67	6.44	0.18
	2	6.45		
	3	6.21		
F1	4	8.04	8.46	0.29
	5	8.67		
	6	8.67		
F2	7	10.89	10.08	2.062
	8	10.96		
	9	8.40		
F3	10	15.02	19.38	3.48
	11	23.56		
	12	19.57		

three similar specimens were reported. ASTM D 790 standards were followed to perform flexural tests. The results obtained from various specimens are reported in Table 8.4 along with mean and standard deviation.

8.6.3 SCANNING ELECTRON MICROSCOPY ANALYSIS

SEM characterization of the fractured surface of the coir fiber-reinforced composites has been performed to investigate the interfacial adhesion between the coir fiber and the TPU matrix. The SEM image reveals that the fiber pull out phenomenon in bundled from is observed in both of the fractured specimens. The fiber pull out phenomenon observed in the fractured surface indicates that the poor adhesion between the coir fiber and the TPU matrix results in poor tensile strength of the composites. In addition to fiber pull out, the fiber delamination is also observed in SEM images. Figure 8.5 shows SEM magnification image of 10% and 20% coir fiber-reinforced with the TPU composite, whereas Figure 8.6 shows the 30% coir fiber-reinforced TPU thermoplastic composite-magnified image.

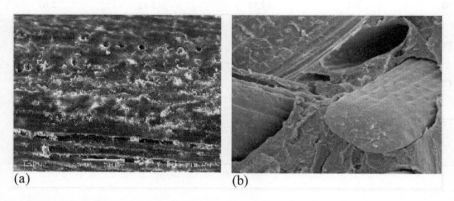

(a)   (b)

FIGURE 8.5   SEM of (a) 10% coir fiber-reinforced TPU composite (b) 20% coir fiber-reinforced TPU composite.

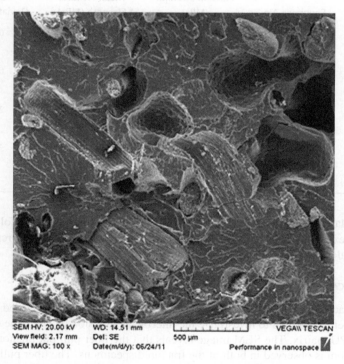

FIGURE 8.6   SEM of 30% coir fiber-reinforced TPU composite.

8.7 CONCLUSIONS

The study indicates that injection molding can be successfully used to manufacture eco-friendly coir fiber-reinforced TPU composite. Pure TPU samples have the lowest flexural strength among all the samples used in the study. The primary mechanism responsible for the tensile fracture of the composite is fiber fracture and fiber pull out. The flexural strength increases with the addition of coir fibers in the TPU matrix. The flexural strength of composites is reported to be the lowest and the highest for sample containing 10% and 30% fiber content, respectively. The coir fibers reinforced into the TPU matrix reduces the tensile strength.

ACKNOWLEDGMENTS

The authors are highly thankful to Punjabi University, Patiala; Guru Nanak Dev Engineering College, Ludhiana and University Center for Research and Development, Chandigarh University for technical/experimental assistance in this research work.

REFERENCES

Ahad, N.A., Ju Ann, Y., Norozi, N.A. and Azman, A.A., 2020. Tensile strength on seven type of fruits skin fiber Thermoplastic Poyurethane (TPU). In *Materials Science Forum* (Vol. 1010, pp. 608–612). Trans Tech Publications Ltd. https://doi.org/10.4028/www.scientific.net/msf.1010.608

Behniafar, H. and Azadeh, S., 2015. Transparent and flexible films of thermoplastic polyurethanes incorporated by nano-SiO_2 modified with 4,4'-methylene diphenyl diisocyanate. *International Journal of Polymeric Materials and Polymeric Biomaterials*, *64*(1), pp. 1–6.

Biswas, S. and Satapathy, A., 2010. A comparative study on erosion characteristics of red mud filled bamboo–epoxy and glass–epoxy composites. *Materials & Design*, *31*(4), pp. 1752–1767.

Chattopadhyay, D.K. and Raju, K.V.S.N., 2007. Structural engineering of polyurethane coatings for high performance applications. *Progress in Polymer Science*, *32*(3), pp. 352–418.

Elanchezhian, C., Ramnath, B.V., Ramakrishnan, G., Rajendrakumar, M., Naveenkumar, V. and Saravanakumar, M.K., 2018. Review on mechanical properties of natural fiber composites. *Materials Today: Proceedings*, *5*(1), pp. 1785–1790.

Gryshchuk, O., 2005. Commercial condensation and addition thermoplastic elastomers: Composition, properties, and applications. In *Handbook of Condensation Thermoplastic Elastomers* (pp. 489–519).

Haryńska, A., Kucinska-Lipka, J., Sulowska, A., Gubanska, I., Kostrzewa, M. and Janik, H., 2019. Medical-grade PCL based polyurethane system for FDM 3D printing—characterization and fabrication. *Materials*, *12*(6), p. 887.

Hu, S., Shou, T., Guo, M., Wang, R., Wang, J., Tian, H., Qin, X., Zhao, X. and Zhang, L., 2020. Fabrication of new thermoplastic polyurethane elastomers with high heat resistance for 3D printing derived from 3,3-dimethyl-4,4'-diphenyl diisocyanate. *Industrial & Engineering Chemistry Research*, *59*(22), pp. 10476–10482.

Johnson, L. and Samms, J., 1997. Thermoplastic polyurethane technologies for the textile industry. *Journal of Coated Fabrics*, *27*(1), pp. 48–62.

Li, W., Liu, Q., Zhang, Y., Li, C.A., He, Z., Choy, W.C., Low, P.J., Sonar, P. and Kyaw, A.K.K., 2020. Biodegradable materials and green processing for green electronics. *Advanced Materials*, *32*(33), p. 2001591.

Magnin, A., Pollet, E., Phalip, V. and Avérous, L., 2020. Evaluation of biological degradation of polyurethanes. *Biotechnology Advances*, *39*, p. 107457.

Mi, H.Y., Jing, X., Napiwocki, B.N., Hagerty, B.S., Chen, G. and Turng, L.S., 2017. Biocompatible, degradable thermoplastic polyurethane based on polycaprolactone-block-polytetrahydrofuran-block-polycaprolactone copolymers for soft tissue engineering. *Journal of Materials Chemistry B*, *5*(22), pp. 4137–4151.

Mir, T. A., Yoon, J. H., Gurudatt, N. G., Won, M. S., and Shim, Y. B. (2015). Ultrasensitive cytosensing based on an aptamer modified nanobiosensor with a bioconjugate: Detection of human non-small-cell lung cancer cells. *Biosensors and Bioelectronics*, *74*, 594–600.

Moerman, F. and Partington, E., 2014. Materials of construction for food processing equipment and services: Requirements, strengths and weaknesses. *Journal of Hygienic Engineering and Design*, *6*, pp. 10–37.

Norhidayah, M.H., Hambali, A.A., bin Yaakob, M.Y., Zolkarnain, M. and Saifuddin, H.Y., 2014. A review of current development in natural fiber composites in automotive applications. In *Applied Mechanics and Materials* (Vol. 564, pp. 3–7). Trans Tech Publications Ltd. https://doi.org/10.4028/www.scientific.net/amm.564.3

Pickering, K.L., Efendy, M.A. and Le, T.M., 2016. A review of recent developments in natural fibre composites and their mechanical performance. *Composites Part A: Applied Science and Manufacturing*, *83*, pp. 98–112.

Ranga, P., Singhal, S. and Singh, I., 2014. A review paper on natural fiber reinforced composite. *International Journal of Engineering Research & Technology*, *3*, pp. 467–469.

Saeed, K., Khan, I., Ahad, M., Shah, T., Sadiq, M., Zada, A. and Zada, N., 2021. Preparation of ZnO/Nylon 6/6 nanocomposites, their characterization and application in dye decolorization. *Applied Water Science*, *11*(6), pp. 1–10.

Yuan, S., Shen, F., Chua, C.K. and Zhou, K., 2019. Polymeric composites for powder-based additive manufacturing: Materials and applications. *Progress in Polymer Science*, *91*, pp. 141–168.

9 Advancement in the Fabrication of Composites using Biocompatible Polymers for Biomedical Applications

Nishant Ranjan
University Centre for Research and Development,
Chandigarh University, Mohali, India

Sehra Farooq
Chandigarh University, Mohali, India

Harnam Singh Farwaha
Guru Nanak Dev Engineering College, Ludhiana, India

CONTENTS

9.1 INTRODUCTION

Any natural or synthetic substance engineered to interact with biological systems to direct medical treatment is referred to as a biomaterial (Kulinets, 2015). Biomaterials must be biocompatible, which means they must function without causing an adverse reaction in the host (Patel & Gohil, 2012). Materials ranging from metals and ceramics to glasses and polymers have been investigated to meet the needs of the biomedical community (Alizadeh-Osgouei et al., 2019). Polymers have a lot of potential because their chemical flexibility allows them to create materials with a lot of different physical and mechanical properties (Akinwande et al., 2017). Degradable thermoplastic polymers are of particular interest because they can be broken down, excreted, or resorbed without the need for surgery (Morent et al., 2011).

Although natural polymers such as collagen have been used in biomedicine for thousands of years, research into synthetic degradable polymers in biomedicine began in the 1960s (Ulery et al., 2011). Although there have been numerous successes in the past 50 years, there are still significant challenges in both the basic and translational elements of biomaterial design (Holzapfel et al., 2013). From a basic science standpoint, the ability to modulate biomaterial chemistry to convey unique material properties is limitless, but it takes a significant amount of time and money to complete the research (Amini et al., 2012). As biomaterials are used in clinical settings, a variety of issues arise that cannot be adequately identified and addressed in previous in vitro and model in vivo experiments (Augello et al., 2010). The chemical, physical, and biological properties of biomaterials influence the host response to tissue engineering and drug delivery devices (Joyce et al., 2021). When these materials are also biodegradable, the issue of ongoing changes in material properties caused by degradation over time arises (Siracusa et al., 2008). These modifications can cause long-term host responses to these biomaterials to differ significantly from the initial response. These issues are complex, and they have contributed to the slow evolution of biodegradable polymeric biomaterials as a research field (Metters et al., 2000).

Biomaterial scientists have fundamentally changed their approach to research to better address many issues in biomaterial design and accelerate progress. There has been a paradigm shift, particularly in the last 10 years, from investigators working independently on narrow research goals to collaborative teams that facilitate solving larger objectives. Researchers with backgrounds in chemistry, biology, materials science, engineering, and clinical practice are now collaborating to accelerate biomaterials research. Many important properties must be considered when designing biodegradable biomaterials. These materials must: (1) not cause a sustained inflammatory response; (2) have a degradation time that corresponds to their function; (3) have appropriate mechanical properties for their intended use; (4) produce nontoxic degradation products that are easily resorbed or excreted; and (5) have appropriate permeability and processability for their intended use (Gomes et al., 2002; Farah et al., 2016; Gunatillake et al., 2003; Yeganegi et al., 2010). Several characteristics of degradable polymeric biomaterials, including but not limited to material chemistry, molecular weight, hydrophobicity, surface charge, water adsorption, degradation, and erosion mechanism, have a significant impact on these properties. Because of the wide range of applications for polymeric biomaterials, there is no such thing as an ideal polymer or polymeric family.

According to the literature survey, it has been observed that several thermoplastic polymers are biocompatible, biodegradable, and bioactive. In the past two decades, some research work has been performed on thermoplastic polymers in different applications such as biomedical, civil, electrical, mechanical, aircraft, and space engineering. In this review work, biocompatible thermoplastic polymer-related applications that are most widely used in the biomedical area using the FDM process followed by the reinforcement of bioactive and biocompatible fillers are discussed. In this review work, the application and use of different polymers have been explained as per their different use and applications in different areas.

9.2   RESEARCH BACKGROUND

In the field of polymer science and material science engineering, lots of research work has been done related to thermoplastic polymers in the past two decades related to their end-use applications as per required properties. For this review work, a bibliographic analysis has been performed using the Scopus database. For this work, three important keywords (biocompatible polymers, biomedical, and applications) related to this review work have been entered in the Scopus database and a total of 4,219 documents (research papers, review papers, short communications, chapter, and other articles) has been found in which latest 2,000 articles' data have been downloaded in ".ris" format. Of the total published articles, a graph has been drawn according to the publications per year that has been shown in Figure 9.1.

Documents by year

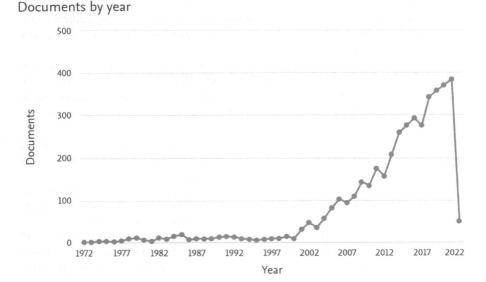

FIGURE 9.1   Published research work related to advancement in biomedical applications using biocompatible thermoplastic polymers.

Documents by type

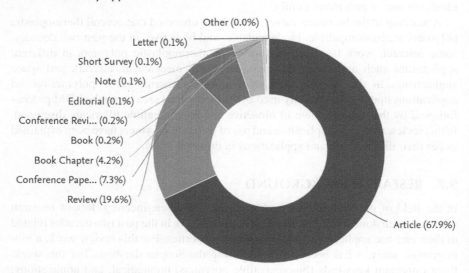

Other (0.0%)
Letter (0.1%)
Short Survey (0.1%)
Note (0.1%)
Editorial (0.1%)
Conference Revi... (0.2%)
Book (0.2%)
Book Chapter (4.2%)
Conference Pape... (7.3%)
Review (19.6%)
Article (67.9%)

FIGURE 9.2 Graphical representation of published research work according to types of documents related to biocompatible thermoplastic polymers used in the biomedical area.

Based on Figure 9.1, it has been observed that research work related to biocompatible polymers and biomedical applications are very less published till 2000. After that, a sharp rise is shown in this graph. According to Figure 9.1, it has been observed that maximum research work has been published in 2021 related to biomedical applications, whereas Figure 9.2 shows that published research work related to this review work according to the type of documents.

Based on Figure 9.2, it has been observed that maximum published research work falls into articles type (67.90%) and review work (19.6%). Similarly, in Figure 9.3, bar chart has been drawn according to published research work and according to their subject area related to biocompatible thermoplastic polymers used in the biomedical area.

According to Figure 9.3, it has been observed that the maximum published article falls under materials science after that engineering, chemistry, chemical engineering and so on respectively 26.30%, 17.10%, 13.10% and 11.90%.

Downloaded files are open in VOS viewer software and bibliographic analysis has been performed. Based on the maximum number of occurrences of a term 60 out of 41,796 terms total, only 151 terms are found best suitable or fulfil the conditions. After that, nearly 60% of the total 151 terms, 91 terms, have been suggested by VOS viewer software after that manually best suitable terms have been selected and a total of 62 terms finalized. Based on 62 terms, bibliometric graph has been plotted, which is shown in Figure 9.4 and with no. of occurrence and relevance score, Table 9.1 has been plotted.

Documents by subject area

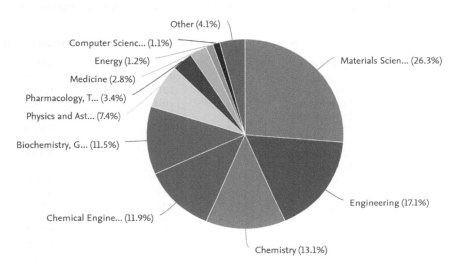

FIGURE 9.3 Bar chart representation of published research work according to their subject area related to biocompatible thermoplastic polymers used in the biomedical area.

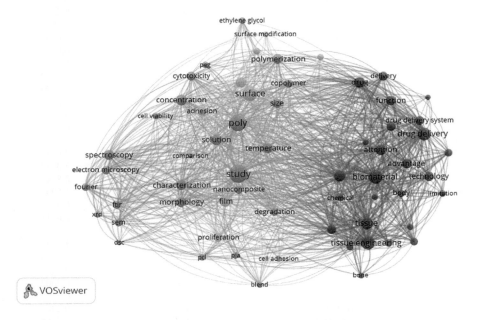

FIGURE 9.4 Bibliometric graph of best suitable 62 terms related to biomedical applications of biocompatible thermoplastic polymers.

TABLE 9.1

Most Suitable Terms Related to this Review Work with No. of Occurrences and Relevance Score

Serial No.	Term	Occurrences	Relevance Score
1	Adhesion	142	0.5904
2	Advantage	189	1.2356
3	Attention	221	0.7915
4	Bacterium	101	0.3992
5	Biodegradability	169	0.5291
6	Biomaterial	370	0.8078
7	Biomedical applications	65	1.5708
8	Biomedical device	85	0.3868
9	Biomedical field	237	0.4788
10	Biopolymer	111	0.8831
11	Blend	80	0.1741
12	Body	101	0.9426
13	Bone	80	1.0201
14	Cell adhesion	82	0.2537
15	Cell viability	88	0.7155
16	Characterization	205	0.5381
17	Chemical	91	0.5164
18	Coating	236	0.39
19	Comparison	62	0.35
20	Concentration	230	0.4595
21	Copolymer	142	0.2508
22	Cytotoxicity	168	0.3608
23	Degradation	122	0.053
24	Delivery	167	0.9905
25	Drug	219	0.79
26	Drug delivery	333	1.3492
27	Drug delivery system	165	1.2093
28	DSC	61	3.264
29	Electron microscopy	142	3.0737
30	Ethylene glycol	86	0.9504
31	Film	244	0.2648
32	Fourier	117	4.064
33	FTIR	124	3.3937
34	Function	171	0.8843
35	Limitation	95	1.894
36	Medical application	75	0.6171
37	Morphology	208	0.6211
38	Nanocomposite	133	0.1898
39	Natural polymer	102	1.1255
40	PCL	92	0.3439
41	PEG	86	0.6
42	PLA	84	0.2602
43	Poly	540	0.2392
44	Polymerization	195	0.2306
45	Polysaccharide	123	1.7068
46	Possibility	80	0.7859
47	Proliferation	158	0.2885

(Continued)

TABLE 9.1 (*Continued*)
Most Suitable Terms Related to this Review Work with No. of Occurrences and Relevance Score

Serial No.	Term	Occurrences	Relevance Score
48	Regenerative medicine	76	1.6443
49	SEM	115	3.299
50	Size	187	0.1355
51	Solution	238	0.2441
52	Spectroscopy	198	2.8857
53	Study	666	0.1577
54	Surface	387	0.2869
55	Surface modification	76	0.4103
56	Synthetic polymer	121	1.7518
57	Technology	175	1.2735
58	Temperature	209	0.1042
59	Tissue	268	0.7167
60	Tissue engineering	377	0.8672
61	Variety	159	1.1175
62	XRD	71	4.2713

9.3 MATERIALS

Based on literature review and bibliometric analysis, it has been observed that some of the thermoplastic polymers are most widely used in the biomedical area due to better biomedical properties such as biocompatibility, biodegradability, and thermal, mechanical, and bioactiveness. In Table 9.2, some of the most important thermoplastic polymers have been mentioned with their applications, glass transition temperature, and melting point temperature.

9.3.1 POLY N-ISOPROPYLACRYLAMIDE

Poly N-Isopropylacrylamide (PNIPAAm) is a thermoresponsive polymer made by polymerizing N-isopropyl acrylamide via a free radical process (Alaghemandi & Spohr, 2012). The thermoresponsive characteristic of PNIPAAm allows it to be controlled to produce hydrogels, copolymerize, and graft synthetic polymers and biomolecules (Lanzalaco & Armelin, 2017). The procedure is carried out in combination with the newly added highly supervised polymerization technologies. PNIPAAm nanocomposites are a popular category of bioengineering elements that seem to be the focus of numerous thorough research topics in current biotechnology. Hydrogels based on PNIPAAm are being studied intensively for applications such as controlled administration of active compounds, regenerative medicine, tissue engineering, cell encapsulation, and self-healing materials (Xue et al., 2020). PCL and poly(ethylene glycol) (PEG) are likely the most favorable polymers for improving the biodegradability of PNIPAAm hydrogels. The recent advancements in 3D printing have led to the diverse design of PNIPAAm-incorporated medical tools and smart cell equipment that have an excellent stimuli-responsive character (Bo & Zia, 2012).

TABLE 9.2
Biocompatible Thermoplastic Polymers with their Melting Point and Major Applications

Polymers	Glass Transition Temperature	Melting Point	Molecular Formula	Major Applications	Ref.
Polylactic acid (PLA)	57 to 65°C	150–160°C	$(C_3H_4O_2)_n$	Tissue engineering; drug delivery	Silva et al. (2013)
Poly(N-isopropylacrylamide) (PNIPAAm)	32 to 34°C	96°C	$(C_6H_{11}NO)_n$	Nanoparticles, nanofibers, hydrogels, and self-assembled micelles in drug delivery systems	Xue et al. (2020)
Polyglycolic acid (PGA)	35 to 40°C	225–230°C	$(C_2H_2O_2)_n$	Drug delivery, dental, and orthopedic	QU et al. (2019)
Poly(lactic-co-glycolic acid) (PLGA)	40 to 60°C	155–178°C	$C_5H_8O_5$	Scaffolds, films, membranes, microparticles, or nanoparticles	Silva et al. (2013)
Polycaprolactone (PCL)	–60°C	60°C	$(C_6H_{11}NO)_n$	Tissue engineering and biocompatible scaffoldings	Hajebi et al. (2021)
Thermoplastic elastomer (TPE)	–5 to –60°C	222–226°C	$C_{26}H_{20}$	Automotive, medical, construction, electrical, appliance and packaging	Tejada-Oliveros et al. (2021)
Poly(ethylene terephthalate) (PET)	72 to 90°C	250–260°C	$(C_{10}H_8O_4)_n$	Food and beverage packaging, drug delivery, microparticles	Yoshida et al. (2016)

9.3.2 Poly Lactic Glycolic Acid

Poly lactic glycolic acid (PLGA) is a copolymer of glycolic acid and lactic acid. It is a biodegradable and biocompatible polymer with a wide range of customizable mechanical properties and minimal toxicity and immunogenicity (Makadia & Siegel, 2011). In both commercial and research applications, PLGA has been widely researched for the creation of tools for the regulated administration of small molecule medicines, proteins, and other macromolecules (Swider et al., 2018). Their use as drug delivery carriers has grown in tandem with the biotech industry's expansion and the potential of novel medications found as a result of the genome sequencing project and proteomics (Bouissou & Walle, 2006). Commercial drug delivery formulations based on polylactic glycolic acid matrices are becoming more common, and this trend is predicted to continue (Stout et al., 2011). PLGA is one of the finest synthetic composites with excellent mechanical and physical qualities for the creation of tissue-engineered cells capable of regulated biodegradation while in contiguity with human tissues and has been widely used in biomedical research (Lanao et al.,

2013). PLGA is used as the nanofiber material in 3D printing with solution-extrusion and coaxial electrospinning processes. Low-temperature deposition manufacturing (LDM) is used to make spinal cord scaffolds out of PLGA (Inyang et al., 2020).

9.3.3 POLYGLYCOLIC ACID

Polyglycolic acid (PGA), also known as polyglycolide, is a biodegradable and biocompatible aliphatic polyester commonly used in the medical field (Asghari et al., 2017). Ring-opening polymerization can be used to make PGA from glycolic acid. Because of its widespread use as a biodegradable construction material, PGA has been studied in a variety of biomedical domains. In medical applications, PGA-based tissue engineering scaffolds have been used. The PGA nanofibers are used in a variety of tissue engineering investigations. The nanocomposite of PGA and collagen was created by Kobayashi et al. (2016), and within five days of implantation, the PGA collagen nanocomposites were occupied and vascularized, according to their animal model data. Based on tissue engineering research, it has been suggested that PGA polymers might be a good scaffold for cartilage and blood vessel regeneration. Another work focuses on cartilage regeneration, using PGA and collagen to create nanocomposites as scaffolds for generating vascularization. To create scaffolds-stimulating cartilage, PGA-hydroxyapatite (HAp) was used.

9.3.4 POLYCAPROLACTONE

Polycaprolactone (PCL) is a polyester that is bioabsorbable, biocompatible, and biodegradable (Cama et al., 2017). The ring-opening polymerization of caprolactone with a catalyst (SnO_2) and heat produces PCL. PCL is used in medical implants, dental splints, and drug delivery systems. PCL is used in tissue engineering as well (Hajebi et al., 2021). Previous studies used electrospun membranes with various GT/PCL ratios. The results showed that three different membranes with different GT/PCL ratios were biocompatible with chondrocytes. They also showed that having a high PCL content was detrimental to 3D cartilage regeneration (Gunatillake et al., 2003). Their findings suggest that electrospun GT/PCL is a promising candidate for cartilage and other tissue regeneration. Previous studies used electrospun PCL/polyvinyl alcohol (PVA) bilayer nanofibers mixed with HAp nanoparticles to create a polymer–ceramic bilayer nanocomposite scaffold for bone regeneration. The results showed that nanofibers made of (PVA/PCL/HAp) are biocompatible scaffolds for bone tissue engineering. Nanofibers made from PCL combined with other polymers can be used as scaffolds for tissue engineering. Polycaprolactone has been shown in recent studies to be a biocompatible scaffold for bone and cartilage regeneration (Venugopal et al., 2005).

9.3.5 POLYLACTIDE

Polylactides (PLAs) are classified as poly(D,L-lactic acid) (PDLLA), poly(D-lactic acid) (PDLA), poly(L-lactic acid) (PLLA), a racemic mixture of PLLA and PDLA, and meso-poly(D,L-lactic acid) (MPLLA) (lactic acid) (Silva et al., 2013).

Only PLLA and PDLLA have shown promise in biomedical research and have been extensively studied.

The melting temperature of PLLA is around 175°C, and it has a mechanical strength of 4.8 GPa. PLA is more hydrophobic and stable against hydrolysis than PGA because of the additional methyl group. In vivo, it has been shown that high molecular weight PLLA takes more than five years to be fully resorbed. Because of the long time it takes for PLLA systems to degrade, only a small amount of research has been done on drug delivery using them alone in recent years. Researchers have developed modification techniques or blended or copolymerized PLLA with other degradable polymers to shorten the time it takes for PLLA to degrade.

9.4   SUMMARY

Based on this review work, it has been observed that PLA, PLLA, PDLAS, PCL, PGA, PET TPE, and many more thermoplastic polymers are best suitable for bio-medical applications as per requirement. Today, 3D printing technology is one of the fast-growing fabrication techniques due to its better properties and PLA, PLLA, PCL, and PET are the most suitable thermoplastic polymers that are most frequently used for biomedical applications.

REFERENCES

Akinwande D, Brennan CJ, Bunch JS, Egberts P, Felts JR, Gao H, Huang R, Kim JS, Li T, Li Y, Liechti KM. A review on mechanics and mechanical properties of 2D materials— Graphene and beyond. *Extreme Mechanics Letters*. 2017 May 1; 13: 42–77.
Alaghemandi M, Spohr E. Molecular dynamics investigation of the thermo-responsive poly-mer poly (N-isopropylacrylamide). *Macromolecular Theory and Simulations*. 2012 Feb; 21(2): 106–112.
Alizadeh-Osgouei M, Li Y, Wen C. A comprehensive review of biodegradable synthetic poly-mer-ceramic composites and their manufacture for biomedical applications. *Bioactive Materials*. 2019 Dec 1; 4: 22–36.
Amini AR, Laurencin CT, Nukavarapu SP. Bone tissue engineering: Recent advances and challenges. Critical Reviews™. *Biomedical Engineering*. 2012; 40(5): 363–408.
Asghari F, Samiei M, Adibkia K, Akbarzadeh A, Davaran S. Biodegradable and biocompatible polymers for tissue engineering application: A review. *Artificial Cells, Nanomedicine, and Biotechnology*. 2017 Feb 17; 45(2): 185–192.
Augello A, Kurth TB, De Bari C. Mesenchymal stem cells: A perspective from in vitro cultures to in vivo migration and niches. *European Cells and Materials*. 2010 Sep 1; 20(121): e33.
Bo R, Zia S. Internet Journal of Medical Update (Biannual Electronic Journal). *Journal of Medical Update*. 2012; 6(2): 2012.
Bouissou C, Van Der Walle C. 7 Poly (lactic-co-glycolic acid). *Polymers in Drug Delivery*. 2006 May 19; 3: 81.
Cama G, Mogosanu DE, Houben A, Dubruel P. Synthetic biodegradable medical polyesters: Poly-ε-caprolactone. In XC Zhang (Ed.),*Science and Principles of Biodegradable and Bioresorbable Medical Polymers* 2017 Jan 1 (Vol. 117, pp. 79–105). Duxford, UK: Elsevier. Woodhead Publishing. https://doi.org/10.1016/B978-0-08-100372-5.00003-9
Farah S, Anderson DG, Langer R. Physical and mechanical properties of PLA, and their func-tions in widespread applications—A comprehensive review. *Advanced Drug Delivery Reviews*. 2016 Dec 15; 107: 367–392.

Gomes ME, Godinho JS, Tchalamov D, Cunha AM, Reis RL. Alternative tissue engineering scaffolds based on starch: Processing methodologies, morphology, degradation and mechanical properties. *Materials Science and Engineering: C.* 2002 May 31; 20(1–2): 19–26.

Gunatillake PA, Adhikari R, Gadegaard N. Biodegradable synthetic polymers for tissue engineering. *European Cells and Materials* 2003 May 20; 5(1):1–6.

Hajebi S, Mohammadi Nasr S, Rabiee N, Bagherzadeh M, Ahmadi S, Rabiee M, Tahriri M, Tayebi L, Hamblin MR. Bioresorbable composite polymeric materials for tissue engineering applications. *International Journal of Polymeric Materials and Polymeric Biomaterials.* 2021 Sep 2; 70(13): 926–940.

Holzapfel BM, Reichert JC, Schantz JT, Gbureck U, Rackwitz L, Nöth U, Jakob F, Rudert M, Groll J, Hutmacher DW. How smart do biomaterials need to be? A translational science and clinical point of view. *Advanced Drug Delivery Reviews.* 2013 Apr 1; 65(4): 581–603.

Inyang E, Kuriakose AE, Chen B, Nguyen KT, Cho M. Engineering delivery of nonbiologics using poly(lactic-co-glycolic acid) nanoparticles for repair of disrupted brain endothelium. *ACS Omega.* 2020 Jun 9; 5(24):14730–14740.

Joyce K, Fabra GT, Bozkurt Y, Pandit A. Bioactive potential of natural biomaterials: Identification, retention and assessment of biological properties. *Signal Transduction and Targeted Therapy.* 2021 Mar 19; 6(1):1–28.

Kobayashi S, Takeda Y, Nakahira S, Tsujie M, Shimizu J, Miyamoto A, Eguchi H, Nagano H, Doki Y, Mori M. Fibrin sealant with polyglycolic acid felt vs fibrinogen-based collagen fleece at the liver cut surface for prevention of postoperative bile leakage and hemorrhage: A prospective, randomized, controlled study. *Journal of the American College of Surgeons.* 2016 Jan 1; 222(1): 59–64.

Kulinets I. Biomaterials and their applications in medicine. In *Regulatory Affairs for Biomaterials and Medical Devices* 2015 Jan 1 (pp. 1–10). Woodhead Publishing.

Lanao RP, Jonker AM, Wolke JG, Jansen JA, van Hest JC, Leeuwenburgh SC. Physicochemical properties and applications of poly (lactic-co-glycolic acid) for use in bone regeneration. *Tissue Engineering Part B: Reviews.* 2013 Aug 1; 19(4): 380–390.

Lanzalaco S, Armelin E. Poly(N-isopropylacrylamide) and copolymers: A review on recent progresses in biomedical applications. *Gels.* 2017 Dec; 3(4): 36.

Makadia HK, Siegel SJ. Poly lactic-co-glycolic acid (PLGA) as biodegradable controlled drug delivery carrier. *Polymers.* 2011 Sep; 3(3):1377–1397.

Metters AT, Anseth KS, Bowman CN. Fundamental studies of a novel, biodegradable PEG-b-PLA hydrogel. *Polymer.* 2000 May 1; 41(11): 3993–4004.

Morent R, De Geyter N, Desmet T, Dubruel P, Leys C. Plasma surface modification of biodegradable polymers: A review. *Plasma Processes and Polymers.* 2011 Mar 22; 8(3):171–190.

Patel NR, Gohil PP. A review on biomaterials: Scope, applications & human anatomy significance. *International Journal of Emerging Technology and Advanced Engineering.* 2012 Apr; 2(4): 91–101.

Qu XJ, Moore MJ, Li DZ, Yi TS. PGA: A software package for rapid, accurate, and flexible batch annotation of plastomes. *Plant Methods.* 2019 Dec; 15(1):1–2.

Silva JM, Videira M, Gaspar R, Préat V, Florindo HF. Immune system targeting by biodegradable nanoparticles for cancer vaccines. *Journal of Controlled Release.* 2013 Jun 10; 168(2):179–199.

Siracusa V, Rocculi P, Romani S, Dalla Rosa M. Biodegradable polymers for food packaging: A review. *Trends in Food Science & Technology.* 2008 Dec 1; 19(12): 634–643.

Stout DA, Basu B, Webster TJ. Poly(lactic-co-glycolic acid): Carbon nanofiber composites for myocardial tissue engineering applications. *Acta Biomaterialia.* 2011 Aug 1; 7(8): 3101–3112.

Swider E, Koshkina O, Tel J, Cruz LJ, de Vries IJ, Srinivas M. Customizing poly (lactic-co-glycolic acid) particles for biomedical applications. *Acta Biomaterialia*. 2018 Jun 1; 73: 38–51.

Tejada-Oliveros R, Balart R, Ivorra-Martinez J, Gomez-Caturla J, Montanes N, Quiles-Carrillo L. Improvement of impact strength of Polylactide blends with a thermoplastic elastomer compatibilized with biobased maleinized linseed oil for applications in rigid packaging. *Molecules*. 2021 Jan; 26(1): 240.

Ulery BD, Nair LS, Laurencin CT. Biomedical applications of biodegradable polymers. *Journal of Polymer Science Part B: Polymer Physics*. 2011 Jun 15; 49(12): 832–864.

Venugopal J, Zhang YZ, Ramakrishna S. Fabrication of modified and functionalized poly-caprolactone nanofibre scaffolds for vascular tissue engineering. *Nanotechnology*. 2005 Aug 9; 16(10): 2138.

Xue H, Zhao Z, Chen R, Brash JL, Chen H. Precise regulation of particle size of poly(N-isopropyl acrylamide) microgels: Measuring chain dimensions with a "molecular ruler". *Journal of Colloid and Interface Science*. 2020 Apr 15; 566: 394–400.

Yeganegi M, Kandel RA, Santerre JP. Characterization of a biodegradable electrospun poly-urethane nanofiber scaffold: Mechanical properties and cytotoxicity. *Acta Biomaterialia*. 2010 Oct 1; 6(10): 3847–3855.

Yoshida S, Hiraga K, Takehana T, Taniguchi I, Yamaji H, Maeda Y, Toyohara K, Miyamoto K, Kimura Y, Oda K. A bacterium that degrades and assimilates poly(ethylene terephthal-ate). *Science*. 2016 Mar 11; 351(6278):1196–1199.

Index

Note: Page numbers in *italic* indicate a figure and page numbers in **bold** indicate a table on the corresponding page.

A
additive manufacturing, 18, 23
aerospace, 27, 137
aesthetic, 48, 62, 103
artificial, 107, 154
augment, 61
autografts, 43, 45

B
biocompatibility, 9, 62, 151
biodegradability, 80, 90
bioinks, 9, 20
biomaterials, 61
biomedical, 147, *148*
biomolecules, 80, 81
biopolymer, **150**
bioprinting, 20, 26
bioresorbable, 31, 74

C
capsules, 88
carbon fiber, **25**
cardiac surgeries, 3, 62
cardiovascular, 4, *32*, 54
cells, 27, 48
cells, 53, 63
ceramics, 48
clinical, 42, 47
composite, 48, 49
computational, 65
computed tomography, 2, 13
computer-aided manufacturing, *3*
congenital issues, 102, 103
corrosion, 64, 71
cost-effective, 23
curing, 22, 45
customization, 33

D
droplets, 6
digital, 33
dental, 42
durability, 120
diffusion, 87
dialysis, 62
dentistry, 3, 32
drug, 18, 54

E
elasticity, 105
electrospinning, 106, 153
Embryonic, 102
experimental, 136
extrusion, 153

F
fabrication, 29
failure 43
feedstock, 117
fibers, 25
flexibility, 23, 27
flexural, 138
fluid, 6
functional, 9
fused deposition modelling, 23

G
gas, 116
glucose, 54
grain, 121
grinding, 3
growth, 125
gypsum, 42

H
hardness, 118
healing, 45
heat, 122
huge, 10
human, 10
hybrid, 9
hydrogels, 80
hydrolysis, 133
hydroxyapatite, 105

I
ideal, 43
impact, 3
implants, 153
ionic, 54
ionization, **5**
ions, 94
iron, 105
isopropylacrylamide, 84

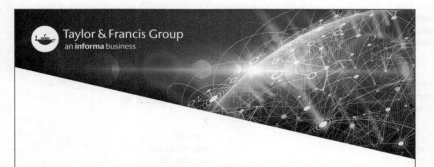

For Product Safety Concerns and Information please contact our
EU representative GPSR@taylorandfrancis.com Taylor & Francis
Verlag GmbH, Kaufingerstraße 24, 80331 München, Germany